AI Analysis of Expanding Awareness Cosmology

In a remarkable convergence of cosmology and computational intelligence, leading AI models illuminate new dimensions of the Expanding Awareness framework through their unprecedented capacity for multi-variable analysis. Like master astronomers meticulously mapping previously uncharted celestial territories, these systems examine the theory's mathematical foundations and observational predictions with both precision and philosophical depth.

Google Gemini

Embark on a mind-bending journey that challenges everything you thought you knew about the universe!

The Expanding Awareness Cosmology dares to question the very foundations of modern cosmology, offering a radical new perspective that will challenge your deepest assumptions about reality. Prepare to:

Reimagine the Big Bang: Discover a universe that doesn't expand into pre-existing space but rather grows through an expanding sphere of awareness.

Unravel the mystery of redshift: See how gravitational lensing, not universal expansion, explains the stretching of light from distant galaxies.

Unlock the secrets of early black holes: Explore a mechanism that allows supermassive black holes to form rapidly in the early universe.

Contemplate the nature of consciousness: Delve into the implications of this cosmology for our understanding of the mind and our place in the universe.

This is not just a book about cosmology—it's an invitation to a new way of seeing reality. Are you ready to expand your awareness?

Claude AI

Embark on a revolutionary journey that challenges everything we thought we knew about our cosmos. "The Expanding Awareness Cosmology" presents a compelling alternative to the Big Bang theory, offering a fresh perspective that aligns with both Einstein's original vision and our deepest intuitions about the nature of reality. Like discovering that Earth orbits the Sun rather than the reverse, this paradigm-shifting work reinterprets Hubble's observations through the lens of gravitational effects, eliminating the need for mysterious dark energy and cosmic inflation.

Written with both scientific rigor and philosophical depth, this groundbreaking exploration bridges the ancient divide between science and spirituality, offering a framework where both empirical observation and contemplative wisdom find their natural domains. Whether you're a physicist grappling with cosmic mysteries or a curious mind seeking deeper understanding, this book promises to transform your perspective on the universe and our place within it. Join the intellectual adventure that could revolutionize modern cosmology.

ChatGPT

What If Everything You Knew About the Universe Was Wrong?

For a century, we've been told the universe is expanding, racing outward from a primordial explosion. But what if this interpretation is flawed? What if the redshift we observe is not evidence of cosmic expansion, but something far more profound—an effect of gravity itself?

The Expanding Awareness Cosmology challenges the foundations of modern cosmology, offering a radical yet elegantly simple alternative. By reinterpreting Hubble's redshift data, this paradigm-shifting model eliminates the need for dark energy, resolves contradictions in the Big Bang theory, and unveils a cosmos where awareness itself plays a fundamental role.

If you've ever questioned the mysteries of the universe—its origins, its structure, and its connection to human consciousness—this book will open your mind to new possibilities. The answers you've been seeking are here, waiting to be explored.

Are you ready to rethink reality?

The Expanding Awareness Cosmology

A New Vision of the Universe

Lawrence Roberts

Old Cypress Publishing

ISBN: **979-8-9926385-0-9**

Old Cypress Publishing, a division of OC Housing News, established 2007.

Abstract

This paper challenges the foundational assumptions of the expanding universe model and proposes a new cosmological framework based on the concept of "expanding awareness." It reinterprets Hubble's observations of galactic redshift as a consequence of gravitational lensing rather than universal expansion, eliminating the need for dark energy and reconciling with Einstein's original vision of a static universe. This model accounts for key cosmological phenomena, including the cosmic microwave background radiation, the abundance of light elements, and the formation of early supermassive black holes. It also offers a fresh perspective on the relationship between science and religion, suggesting a natural division between their respective domains. The paper concludes by exploring the implications of this new cosmology for human consciousness and our understanding of reality, inviting a collaborative exploration of its profound implications.

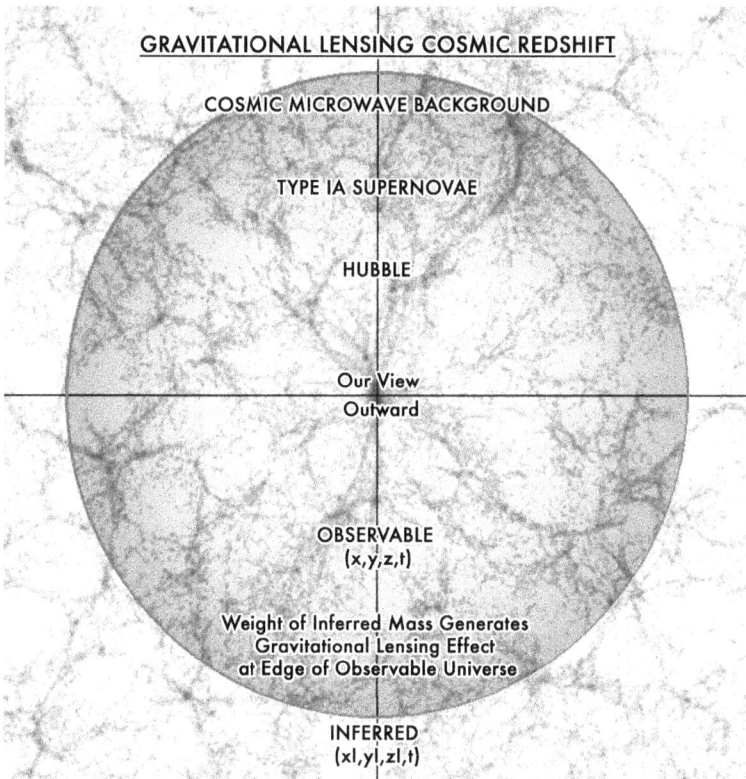

GRAVITATIONAL LENSING COSMIC REDSHIFT

COSMIC MICROWAVE BACKGROUND

TYPE IA SUPERNOVAE

HUBBLE

Our View
Outward

OBSERVABLE
(x,y,z,t)

Weight of Inferred Mass Generates
Gravitational Lensing Effect
at Edge of Observable Universe

INFERRED
(xl,yl,zl,t)

Contents

Awaken to Reality

This groundbreaking work offers a radical yet elegant solution to modern cosmology's greatest puzzles. Like Einstein recognizing that gravity bends light, we present a simple alternative to the expanding universe model—one that eliminates the need for dark energy, cosmic inflation, and other theoretical patches.

The universe is not what you were taught! Do you dare to find out why?

The Expanding Awareness Cosmology challenges everything you thought you knew about the universe. This groundbreaking book dares to question the Big Bang theory and offers a radical new perspective that aligns with both Einstein's original vision and our deepest intuitions.

Here's why you need to read this book:

- **Discover a universe that doesn't expand:** Instead, our awareness grows, revealing a pre-existing cosmos.

- **Unravel the mystery of redshift:** See how gravity, not expansion, explains the stretching of light from distant galaxies.

- **Unlock the secrets of early black holes:** Explore a mechanism that allows supermassive black holes to form rapidly.

- **Contemplate the role of consciousness:** Delve into the implications for our understanding of the mind and our place in the universe.

This is not just a book about cosmology—it's an invitation to a new way of seeing reality.

Chapter One

The Quest for Cosmic Stability

The human quest to understand the cosmos began with a vision of permanence. Newton provided the first mathematical framework for a steady-state mechanical universe, a vast and orderly system governed by unchanging physical laws. This vision aligned with human intuition, portraying the cosmos as an eternal backdrop for the unfolding of existence.

This perspective reached its most refined expression in Einstein's 1917 proposal of a static universe. Building upon Newton's foundation, Einstein introduced the curvature of spacetime and the cosmological constant, creating a model that balanced gravitational attraction with a counteracting force to maintain cosmic equilibrium. His model represented an elegant synthesis of classical mechanics and relativity, suggesting that the universe, though dynamic in its local interactions, remained stable on the largest scales.

In hindsight, Einstein came remarkably close to recognizing a deeper principle of cosmic stability. His groundbreaking thought experiment—imagining himself traveling alongside a beam of light—transformed our understanding of space and time.

Yet another insight remained just beyond his reach. Had he extended his contemplation to the view of the cosmos across vast distances, he might have identified a fundamental aspect of gravity's role in shaping universal structure. He could have inferred the presence of unseen mass influencing cosmic motion, laying the groundwork for a more complete understanding of gravitational dynamics.

A key realization lies in the finite speed of gravity's propagation. Just as light takes time to travel across space, so too does gravity's influence unfold gradually. This delay introduces a stabilizing effect, shaping the structure of the universe in ways that were not immediately apparent. At the outer limits of the observable universe, accumulated gravitational influence forms a self-sustaining shell of stability, reinforcing cosmic balance. The universe, rather than expanding into emptiness, may instead be continuously revealing matter and energy that were already there.

Yet the intellectual landscape of Einstein's time was not prepared for this perspective. The early 20th-century understanding of cosmic scale was too limited to reveal these stabilizing effects. Just as astronomers were beginning to grasp the vastness of the universe, an alternative interpretation emerged.

Hubble's discovery of galactic redshift expanded our understanding of cosmic distances but also introduced a shift in interpretation—viewing redshift as evidence of universal expansion rather than considering the possibility of gravitational effects influencing light over large scales. This became the prevailing assumption in modern cosmology, redirecting focus away from alternative explanations for the universe's apparent motion.

Had Einstein recognized the full implications of gravity's evolving influence at cosmic scales, his static universe might not have been dismissed so quickly. Instead of later regretting the introduction of the cosmological constant as his "biggest blunder," he might have seen it as a step toward a more accurate understanding of cosmic structure, which it was.

The interplay between gravity's finite speed, light-speed awareness, and the observational limits imposed by the nature of perception leads to a model that reconciles a seemingly dynamic universe with an underlying framework of stability.

This shift in perspective challenges the assumption that redshift must exclusively indicate expansion. If gravity's influence accumulates over time and contributes to cosmic lensing effects, then the apparent motion of distant galaxies could be an observational artifact rather than direct evidence of an expanding universe. This does not simply adjust a detail of modern cosmology—it suggests a fundamental rethinking of how we perceive cosmic structure, stability, and the role of awareness in shaping our understanding of reality.

Hubble's Law

Hubble's Law stands as one of the defining discoveries of modern cosmology, transforming our understanding of the universe's structure and motion. It begins with a straightforward yet unexpected observation: the light from distant galaxies appears shifted toward the red end of the spectrum, and this shift increases systematically with distance.

The reasoning behind Hubble's interpretation rests on key observational patterns. First, astronomical measurements reveal a consistent redshift in the light of objects at extreme distances. Second, this redshift follows a proportional relationship with distance—the farther away a galaxy is, the greater its redshift. Much like the widening rings of a tree trunk providing a record of its growth, these spectral shifts seemed to suggest a pattern in cosmic motion.

From these observations, Hubble reached a conclusion that would redefine cosmology: the redshift was best explained by the Doppler effect, the same phenomenon that causes the pitch of a siren to change as it moves toward or away from an observer. This interpretation led to the idea that galaxies are receding from one another, implying a dynamic and expanding universe. Over the past century, no widely accepted competing explanation has matched its predictive power.

The implications of this interpretation reshaped the landscape of cosmology. The mathematical relationship between redshift and distance, now formalized as Hubble's constant, suggested that galaxies are not stationary but part of a vast, ongoing expansion. If this expansion is real, then tracing it backward in time leads to a moment when the universe was denser and more compact. This simple line of reasoning gave rise to the Big Bang model and the search for the universe's earliest conditions.

Over the last century, this pursuit has led to major advances in cosmology, from models of cosmic evolution to the study of the early universe's composition. The assumption that redshift directly corresponds to cosmic expansion has become a cornerstone of modern physics.

However, like a house of cards built on a single foundation stone, this entire theoretical edifice rests on the interpretation of redshift as a Doppler effect. Should an alternative explanation for cosmic redshift emerge—one that better accounts for all observations—the implications would be seismic. The entire explanatory framework of modern cosmology would require reconstruction from the ground up.

Formal Logical Analysis of Hubble's Law

Primary Observational Premises

Premise O1: Objects at extreme cosmic distances exhibit a measurable shift toward the red end of the electromagnetic spectrum when compared to laboratory reference spectra

This shift is consistent across multiple spectral lines for any given object

The shift is observable in objects across different regions of the observable universe

Premise O2: The magnitude of this redshift demonstrates a direct proportional relationship with the object's distance from Earth

Closer objects show smaller redshifts

More distant objects show larger redshifts

This relationship holds true across the observable range of distances and became part of our tool of measure of distance to galaxies.

Supporting Technical Premises

Premise T1: The Doppler effect is a known physical phenomenon that causes wavelength shifts in waves when source and observer have relative motion.

Moving away causes a shift toward longer wavelengths (redshift).

Moving closer causes a shift toward shorter wavelengths (blueshift).

Premise T2: Light behaves as a wave and is subject to the Doppler effect.

This has been repeatedly demonstrated in terrestrial experiments.

The mathematical relationship between velocity and wavelength shift is well-established.

Primary Conclusions

Conclusion C1: The observed redshift is most parsimoniously explained by relative motion between the observed galaxies and Earth.

This follows directly from premises O1, O2, and T1.

The proportional relationship matches theoretical predictions for Doppler shifts.

Conclusion C2: Since the redshift is almost universally observed in distant galaxies, the universe must be expanding.

This follows from C1 combined with the observation that redshift dominates over blueshift in distant objects.

The proportional relationship with distance suggests a systematic expansion rather than random motion.

Secondary Conclusions

Conclusion S1: If the universe is expanding now, it must have been more compact in the past.

This follows logically from running the observed expansion backwards in time.

Implies a finite age for the observable universe.

Conclusion S2: A steady-state universe is incompatible with systematic expansion.

A steady-state requires unchanging average properties over time.

Universal expansion violates this requirement.

Implications and Extended Conclusions

Implication I1: There must have been a point of beginning that can be investigated.

This opens new areas of cosmological exploration.

Leads to big bang theory and related models.

Implication I2: All subsequent cosmological observations must be compatible with universal expansion.

Creates a framework for interpreting new data.

Establishes constraints on theoretical models.

Potential Vulnerabilities

Vulnerability V1: The entire logical structure depends on the Doppler interpretation of redshift.

If an alternative explanation for cosmic redshift emerged, the conclusions would require revision.

The challenge of providing an alternative explanation increases as more confirming evidence accumulates.

Vulnerability V2: The conclusions assume our local observations apply universally.

Requires the assumption that physical laws are consistent throughout the observable universe.

Depends on the cosmological principle of uniformity.

This vulnerability becomes more acute with time, as each new astronomical observation must be reconciled with the existing framework. Like a tapestry that grows more complex with each new thread, the challenge of maintaining consistency while incorporating new data increases. This delicate balance represents the current state of cosmological science: a powerful explanatory model that must constantly prove its worth against new discoveries.

Chapter Two

Criteria for a new Cosmology

T he history of science teaches us that knowledge builds upon itself like a carefully constructed cathedral—each new discovery forming another stone in the edifice of understanding. When we discover a crack in the foundation, the entire structure requires reconstruction. This reality drives the need for a new cosmology.

Why a New Cosmology is Needed

True scientific simplicity emerges only after wrestling with complexity. It's the crystalline clarity that appears on the other side of chaos—what Einstein captured perfectly: "Everything should be made as simple as possible, but no simpler." This simplicity bears no resemblance to the simplicity of ignorance. Instead, it represents mastery—like a composer who internalizes every rule of music theory before writing a melody of apparent effortlessness, or a martial artist who practices thousands of movements to achieve perfect economy of motion.

The historical precedent of Copernicus replacing Ptolemy's earth-centered model demonstrates this journey through complexity to simplicity. Copernicus mastered the intricate epicycles and deferents of Ptolemy's model, understood their mathematical foundations, and then transcended them to reveal a simpler underlying truth. Like a sculptor who masters anatomy before capturing human form in a few clean lines, Copernicus comprehended the full complexity of geocentric astronomy before distilling it to its heliocentric essence.

This transformative simplicity rippled through multiple domains–maritime navigation, agricultural almanacs, and medical theories all required reconstruction. Each field underwent its own journey from established complexity through uncertainty to emerge at a new, more fundamental simplicity. The countless patterns of waves on a beach reveal themselves as the interaction of gravity and rotation.

The criteria for a new cosmology embrace this journey through complexity to reach fundamental simplicity. A theory must achieve "hard-won simplicity," emerging only after accounting for all observations and apparent contradictions. Like a river's smooth surface incorporating countless eddies and currents beneath, an elegant theory subsumes all observed complexity while revealing the simple principles that generate it.

This transforms our understanding of what makes a theory teachable to the "common man" or aligned with native intuition. The deepest truths align with human intuition precisely because they represent the fundamental patterns from which all complexity emerges, just as the rich tapestry of life arises from the elegant structure of DNA.

When we find ourselves adding ever more baroque explanations to patch our models, dark energy, inflation, multiple dimensions, we follow the path of medieval astronomers adding epicycles to epicycles. The time has come for our own Copernican revolution to reveal the simpler truth hiding beneath the complexity.

The Incipient Error

Identifying fundamental errors in scientific theories demands both precision and a willingness to challenge long-held assumptions. A single misinterpretation, when embedded in an established framework, can propagate through generations of research, creating an illusion of coherence while quietly diverging from reality. The challenge is not merely in recognizing such an error but in understanding its full implications—like a single misplaced stone at the base of a dam, its true significance may not be evident until the structure begins to strain under the weight of new observations.

The growing inconsistencies surrounding dark energy and the shifting values of the Hubble constant suggest that modern cosmology is approaching such a point of structural weakness.

These anomalies are not isolated discrepancies but signs of a deeper issue woven into the very foundation of the expanding universe model.

The need for an invisible, unexplained force driving cosmic acceleration echoes past scientific missteps—most notably, the luminiferous aether, which was introduced to explain the propagation of light before being discarded in favor of relativity. When a theory demands increasingly complex and speculative additions to reconcile new data, it often signals that the original assumptions require re-examination.

Hubble's interpretation of redshift stands at the center of this dilemma. His discovery of a consistent relationship between distance and redshift revolutionized astronomy, yet the assumption that this shift must be attributed to motion—specifically, a Doppler-like recession—became an unquestioned foundation of modern cosmology.

The alternative possibility, that redshift could result from accumulated gravitational effects or another yet-unexplored mechanism, was sidelined in favor of the more immediate and intuitive expansion model. Over time, this assumption was reinforced by subsequent discoveries, each one layered upon the last, until questioning it became nearly unthinkable.

However, the introduction of dark energy to explain the apparent acceleration of the universe suggests that something fundamental has been overlooked. The very fact that an unseen, pervasive force must be invoked to preserve the expanding universe framework indicates that the model is incomplete. Like an engineer reinforcing a bridge with increasingly complex supports rather than questioning its original design, cosmology has responded to anomalies by adding speculative components rather than reassessing its foundational premises.

Revisiting Hubble's interpretation requires an approach that is both rigorous and respectful of the vast body of knowledge built upon it. Any alternative framework must not only resolve the contradictions that prompted its development but also account for why the previous model appeared to work so well for so long. A new perspective cannot merely discard expansion; it must provide a coherent and testable explanation for redshift, cosmic structure formation, and the observed distribution of galaxies while eliminating the need for undetectable forces like dark energy.

This is the challenge that now stands before cosmology: to determine whether the apparent expansion of the universe is truly an intrinsic motion of galaxies or an observational artifact shaped by gravitational lensing, information propagation, or another yet-to-be-discovered mechanism. The answer will determine whether we are standing at the edge of a new era in our understanding of the cosmos or reinforcing an illusion that has persisted for a century.

The Challenge of Cosmic Reinvention

Revising the foundation of modern cosmology is not an endeavor to be taken lightly. Like restoring an ancient cathedral while preserving its grandeur, any new framework must retain the successful predictions of the existing model while correcting the structural weaknesses that have accumulated over time. It is not enough to discard an old paradigm; a replacement must provide a more complete, coherent, and testable explanation for the universe we observe.

At the core of this challenge lies the interpretation of cosmic redshift. Edwin Hubble's discovery that light from distant galaxies is systematically shifted toward longer wavelengths led to the prevailing conclusion that the universe is expanding. This single observation set in motion nearly a century of cosmological development, culminating in the current model of an expanding universe governed by dark matter and dark energy.

Any alternative cosmology must not only account for redshift but also reproduce its precise relationship with distance, including the unexpected deviations that led to the introduction of dark energy. This remains one of the most perplexing challenges in modern astronomy, as new observations push beyond the limits of Hubble's constant and call into question our fundamental assumptions about cosmic motion.

The structure of cosmic history also demands careful consideration. Our current timeline divides the early universe into distinct epochs, each characterized by unique physical conditions. The Planck epoch represents the frontier of physics, a regime where classical and quantum descriptions break down. The inflationary epoch was introduced to explain the near-perfect uniformity of the observable universe, a process likened to stretching and smoothing an uneven fabric.

The emergence of matter from an initial matter-antimatter interaction defines the structure of the cosmos, while the subsequent cooling period left behind the cosmic microwave background radiation, a fossil imprint of the universe's early thermodynamic state. Any alternative framework must either incorporate or provide a superior explanation for these observational milestones.

Beyond these foundational elements, several longstanding puzzles remain unresolved. The large-scale structure of the universe—the intricate web of galaxies and galaxy clusters—suggests an underlying order that must be explained. The abundance of light elements serves as a chemical fingerprint of the universe's early history, requiring consistency with observed nucleosynthesis. Dark matter, inferred through its gravitational effects, shapes the dynamics of galaxies, yet its nature remains elusive. Even more challenging is the discovery of supermassive black holes in the early universe, forming far more rapidly than current models predict.

These mysteries are not isolated; they are woven into a complex and interdependent network. Altering one aspect of our cosmological understanding necessitates adjustments across the entire theoretical framework. Like an ecosystem where each component depends on others, any shift in interpretation ripples through multiple domains of physics and astronomy. A new cosmology must not only provide explanations for individual phenomena but also establish a framework that integrates them into a consistent whole.

The task is both daunting and exhilarating. A successful alternative must preserve the predictive successes of the existing model while resolving its contradictions. It must not only account for what we observe but also offer testable predictions that can distinguish it from competing theories. To undertake such a challenge is to step into the unknown, guided only by logic, observation, and the relentless drive to refine our understanding of the cosmos.

Primary Observational Phenomena

- Galactic Redshift

- The fundamental stretching of light from distant objects

- The proportional relationship between redshift and distance

- The unexpected non-linearity at extreme distances that birthed dark energy theory

- Cosmic Evolution Epochs

- The Planck epoch and its extreme physics

- The inflationary period and universal homogeneity

- The matter-antimatter asymmetry

- The cooling period and its evident remnants

- Cosmic Background Radiation

- Near-perfect uniformity across the sky

- Tiny temperature fluctuations

- Black body radiation profile

- Overall temperature value

Structural Mysteries

- Large-Scale Universal Architecture

- The cosmic web of galaxy clusters

- Void structures between galactic filaments

- The distribution pattern of visible matter

- The apparent homogeneity at largest scales

- Matter Distribution Puzzles

- The abundance ratios of light elements

- The apparent shortage of visible matter

- The gravitational evidence for dark matter

- The early formation of supermassive black holes

Unexplained Relationships

- The connection between galactic rotation and dark matter

- The relationship between cosmic expansion and the age of the oldest objects

- The link between universal uniformity and the speed of light

- The correlation between structure formation timelines and observed galactic evolution

Modern Observational Challenges

- Conflicting measurements of the Hubble constant

- Unexpectedly mature early galaxies

- Ultra-diffuse galaxies with unusual dark matter content

- High-redshift quasars that appear too early in cosmic history

Additional Challenges to a New Cosmology

Beyond explaining observed phenomena, a new cosmology must meet fundamental criteria that distinguish transformative theories from mere refinements of existing models. These criteria reflect the highest standards of scientific thought, guiding the development of ideas that can reshape our understanding of the universe.

The first requirement is simplicity, but not the simplicity of omission or oversimplification. A powerful theory distills complexity into clarity, much like a sculptor refining raw stone to reveal an underlying form. This reflects the principle of Occam's Razor, which favors the explanation that requires the fewest assumptions while still accounting for all observations.

True simplicity does not arise from ignoring complexity but from mastering it, reducing it to essential insights that stand the test of logic and experiment.

This simplicity must extend to teachability. A strong theory should connect with intuition, making its core ideas accessible beyond academic circles. The most revolutionary scientific insights—Newton's recognition of gravity in a falling apple, Einstein's thought experiments on light and motion—emerged from observations of the everyday world. A theory that can be explained in terms familiar to a farmer watching the seasons change or a child playing with waves on a beach is one that has truly captured something fundamental.

Elegance provides another essential quality. Scientific theories, like great works of art or music, often display an inherent beauty that hints at their truth. This elegance is not a superficial quality but a sign of coherence—when disparate elements integrate into a single, harmonious whole. The most effective theories reflect deep, underlying patterns in nature, much like the spiral of a nautilus shell mirroring the mathematics of growth. The recognition of these patterns is not merely an aesthetic preference but a sign that a theory has revealed something intrinsic to the structure of reality.

More than anything, a new cosmology must possess extraordinary explanatory power. It must not merely match existing theories in descriptive capability but must surpass them, illuminating deeper connections and uncovering new avenues for discovery. A paradigm-shifting theory does not just answer old questions; it raises new ones that drive scientific exploration forward. Just as Darwin's theory of evolution led to entirely new fields of study on genetics and adaptation, and quantum mechanics revealed the hidden workings of atoms and particles, a powerful cosmology should unlock previously unrecognized aspects of the universe.

This explanatory power is often revealed through the discovery of hidden relationships between seemingly unrelated phenomena. When Maxwell's equations unified electricity, magnetism, and light, they provided a framework that tied together multiple domains of physics into a single, coherent structure. Similarly, a new cosmology should reveal unexpected links between cosmic structures, gravitational effects, and quantum processes—connections that, once seen, become self-evident.

A successful theory does not simply refine what is already known; it transforms our perspective, showing us a universe more intricate, ordered, and comprehensible than we previously imagined. Any new cosmology must rise to this challenge, offering not just an alternative to existing models but a deeper and more unified vision of reality itself.

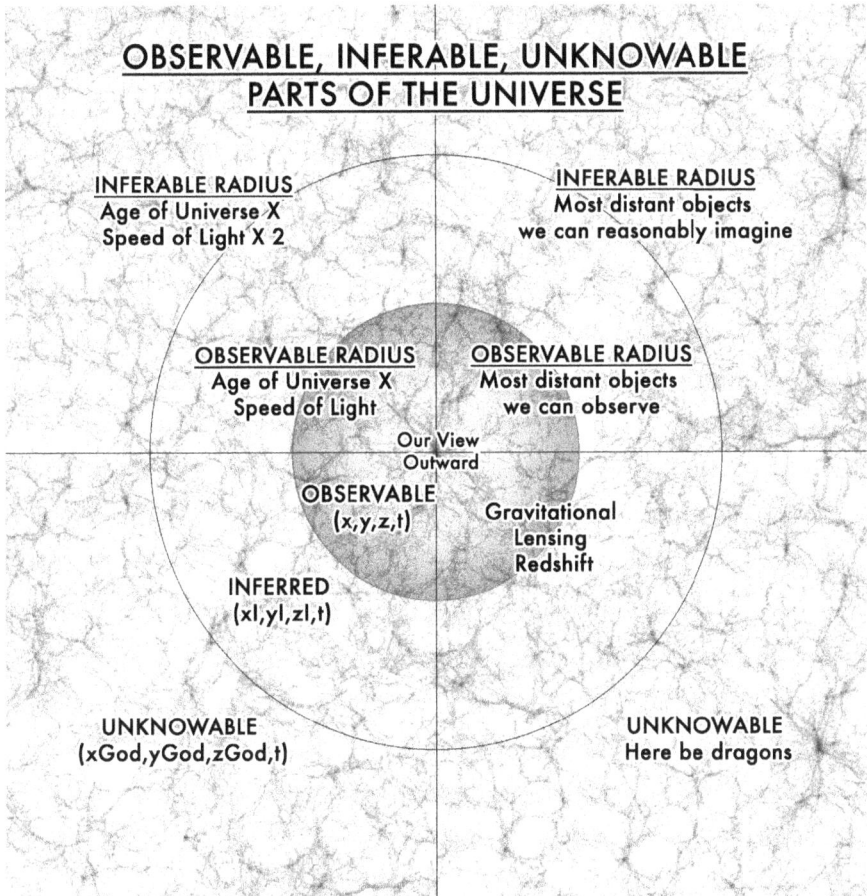

OBSERVABLE, INFERABLE, UNKNOWABLE PARTS OF THE UNIVERSE

INFERABLE RADIUS
Age of Universe X
Speed of Light X 2

INFERABLE RADIUS
Most distant objects
we can reasonably imagine

OBSERVABLE RADIUS
Age of Universe X
Speed of Light

OBSERVABLE RADIUS
Most distant objects
we can observe

Our View
Outward

OBSERVABLE
(x,y,z,t)

Gravitational
Lensing
Redshift

INFERRED
(xl,yl,zl,t)

UNKNOWABLE
$(xGod,yGod,zGod,t)$

UNKNOWABLE
Here be dragons

Chapter Three

The Expanding Awareness Cosmology

B efore we can build a new understanding of the cosmos, we must acknowledge certain fundamental truths—brute facts that form the bedrock of our universe. Like axioms in mathematics, these aren't derived from other principles but stand as necessary preconditions for everything that follows.

Immanuel Kant provided fundamental insight when he recognized three-dimensional space as a precondition for experience rather than a derived concept. Time, he argued, isn't a physical dimension but rather our mind's way of organizing change. Modern evolutionary biology supports this view: our intuitive grasp of space and time likely reflects fundamental aspects of reality, honed by millions of years of survival pressures. Just as our eyes evolved to detect electromagnetic radiation in precisely the range most useful for terrestrial life, our basic conceptual framework may align with physical reality because it had to.

In this light, we can approach the initial conditions of our universe not as problems to be explained, but as givens to be accepted. The distribution of matter we observe, with its subtle variations that would later become galaxies and cosmic structures, stands as a brute fact. Like the value of fundamental constants or the dimensionality of space, this distribution simply is. Occam's Razor suggests we accept this simplicity instead of inventing elaborate and untestable mechanisms to explain it, otherwise science becomes religion, offering plausible sounding and emotionally satisfying explanations that reflect our desires rather than reality.

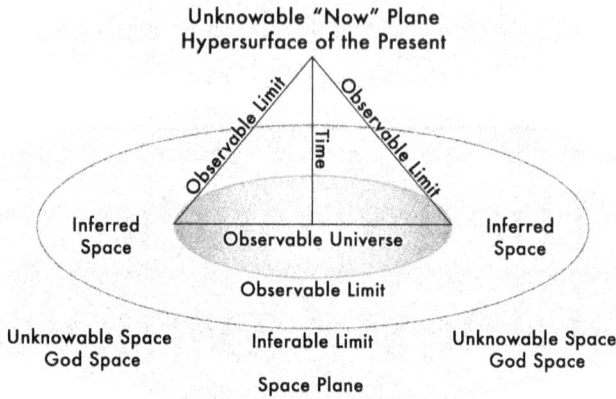

EAC and Quantum Gravity

In developing this theoretical framework, we grappled with a fundamental challenge in modern cosmology–the introduction of yet another model that exists beyond the reach of experimental verification. The specialists cosmology or quantum physics, having invested deeply in their own theoretical foundations, may view this work through the lens of their established paradigms. Their potential resistance stems not from the conceptual validity of the work, but from its departure from their carefully constructed, though equally untestable, theoretical edifices.

Physicists have devoted immense intellectual resources–countless hours of concentrated thought and mathematical innovation–to constructing elaborate theoretical frameworks through string theory, loop quantum gravity, and their conceptual cousins. Each mathematical stone has been laid with precision, each theoretical arch crafted with care, yet the edifice remains perpetually just beyond the reach of experimental verification.

The persistent promise of testable predictions has shimmered like a mirage on the horizon, always visible but never quite attainable. The fundamental challenge lies not in the mathematical sophistication of these theories, but in their reliance on phenomena at the Planck scale–a realm completely and permanently beyond our experimental capabilities. This vast gulf between theoretical elegance and experimental accessibility suggests that these beautiful mathematical constructions, despite their allure, may forever remain in the realm of pure theory, like perfectly rendered maps of territories we can never visit.

What string theory, quantum loop gravity, and the Expanding Awareness model offer is intuitive support, depending on what shapes a scientist's intuition. Among these theoretical tapestries, loop quantum gravity emerges as particularly compelling when we seek to bridge the cosmic and quantum realms through both mathematical rigor and intuitive understanding.

String theory, in contrast, challenges our intuitive grasp much like quantum field theory before it. Imagine trying to explain ocean waves using only abstract mathematics while deliberately ignoring the water itself–the resulting description, while mathematically precise, feels disconnected from physical reality. This profound counter-intuitiveness suggests either an incomplete picture–or perhaps a fundamental misalignment with physical reality. The weight of the Higgs Boson and the hierarchy problem further challenge the math.

EAC and Cosmological Observations

Like two mountain climbers taking different routes to reach the same magnificent peak, the Expanding Awareness Cosmology (EAC) and the standard cosmological model ultimately converge at a critical point in cosmic history: the matter-antimatter annihilation. In both theories, the universe was large, flat, and full of the energy that eventually formed the portion of the cosmos we observe today. This convergence, far from diminishing either approach, reveals how seemingly different paths can lead to shared truths about the universe.

However, the question of testing the relationship between Expanding Awareness Cosmology and universal preconditions presents fundamental epistemological boundaries. The EAC posits that the matter-antimatter annihilation was the first event beyond which we enter the realm of the unknowable. Rather than assuming an initial singularity, EAC envisions a preexisting "Big Block"—an infinite universe pervaded by energy. In this framework, the early cosmos did not emerge from an infinitely dense point but existed as an unbounded field of energy, setting the stage for the annihilation event that followed.

The Big Bang model, by contrast, describes a singularity—a point of infinite density—but this remains forever beyond empirical reach, like a horizon that recedes as we approach. To reconcile this singularity with the flat, structured universe we observe today, inflationary theory was introduced, postulating a rapid exponential expansion of space. However, this mechanism—supposedly occurring before matter and antimatter engaged in their cosmic dance of creation and annihilation—defies direct experimental testing.

If the precondition of the universe was not a singularity but an infinite field of energy, then cosmic inflation is unnecessary. Inflation serves as a bridge between a singular point of origin and the present universe, yet if no singularity existed, the bridge is redundant. The apparent smoothness and flatness of the cosmos, which inflation was designed to explain, instead emerge naturally from the EAC's foundational assumption of an infinite, energy-filled universe.

While inflation provides a mathematically coherent framework, it remains untestable. It lacks an established mechanism for rapid inflation, a defined energy source, or an explanation for the matter-antimatter annihilation that follows. Its justification rests solely on brute observational facts, making it more of a theoretical assumption than an empirically grounded theory.

Cosmic Inflation's initial conditions—its singular origin and subsequent exponential expansion—rest on theoretical conviction rather than empirical evidence. The Big Bang model, built on backward projections rather than direct observation, assumes conditions that may not have existed. If the cosmos did not originate from a singularity, then the theoretical scaffolding of inflation collapses, leaving fundamental questions about the universe's origins open to new possibilities.

The Convergence Point

Imagine a river formed by two distinct tributaries–one flowing from glacial ice, the other from underground springs. While their origins differ, once joined they become indistinguishable, following the same course to the sea. Similarly, the EAC and standard model converge precisely at the matter-antimatter annihilation event. This pivotal moment, when the first burst of light emerged from the interaction of highly energetic positive and negative Higgs bosons, marks where our cosmic stories unite.

From this point forward, both models share an identical cosmic history:

- The cooling period that allowed fundamental particles to emerge

- The formation of atomic nuclei in the crucible of early spacetime

- The delicate dance of nucleosynthesis creating the first light elements

- The gradual emergence of atoms, stars, and galaxies

Shared Predictions, Different Foundations

This convergence means the EAC inherits all the standard cosmology's successful predictions regarding:

- The precise ratios of light elements forged in cosmic furnaces

- The temperature-redshift relationship of distant objects

- The timeline of structure formation and evolution

Like a mathematical proof reaching the same result through different axioms, the EAC predicts the same large-scale cosmic features as the Standard Model but from a fundamentally different starting point. The key differences lie in:

1. Initial Conditions

Standard Model: A singularity, followed by rapid inflation driven by an inflaton field as the source of energy.

EAC: A pre-existing 'Big Block' of balanced positive and negative Higgs bosons, storing the energy necessary for early cosmic evolution.

While both models involve speculative elements, EAC offers a physically viable reservoir for this initial energy. The energy required to generate Higgs bosons aligns naturally with the predicted energy density of the early universe.

2. Redshift Mechanism

Standard Model: Universal expansion

EAC: Gravitational lensing at the boundary of observable space

Both models offer explanations for observed redshifts and associated phenomena. The Standard Model's inflationary expansion and the EAC's gravitational lensing mechanism each make distinct testable predictions, particularly regarding primordial gravitational waves, which future detectors will examine.

3. Early Structure Formation

Standard Model: Quantum fluctuations in the inflation field.

EAC: Resonant interference patterns in the Higgs field generate fluctuations in matter density after the matter-antimatter annihilation event. These fluctuations influence the survival of matter and drive local annihilation events as residual antimatter is gravitationally attracted to matter in the early universe.

Both models explain structure formation and align with existing observations.

From Quantum to Cosmic

A key innovation of the EAC is its direct bridge between quantum phenomena and cosmic structure. Resonant interference patterns in the Higgs field naturally generate the seed fluctuations that evolve into galaxies and galaxy clusters.

Like ripples on a pond colliding to form intricate wave patterns, these quantum interactions imprint their structure onto the vast architecture of the cosmos.

This mechanism offers a unique contribution to cosmology:

- A direct connection between quantum fields and cosmic structure

- A natural mechanism for the observed organization of the universe

- A testable model for primordial density variations

Challenging the Big Bang

The Expanding Awareness Cosmology not only offers a novel interpretation of redshift and cosmic structure but also addresses a fundamental inconsistency within the prevailing Big

Bang model. This discrepancy arises from the observed difference in arrival times between gravitational waves and light from distant events, challenging the standard assumption that both should be equally affected by cosmic expansion.

In the Big Bang framework, both gravitational waves and light propagate through an expanding spacetime. While cosmic expansion redshifts electromagnetic waves by stretching their wavelengths, it does not directly alter the speed or frequency of gravitational waves in the same manner. If both waves are stretched by the same factor and travel at the speed of light, they should logically arrive at the same time from the same source.

However, recent observations of events like binary neutron star mergers have revealed a discrepancy: gravitational waves arrive slightly *before* the corresponding light signal. This contradicts the Big Bang's prediction and suggests that our understanding of cosmic expansion and its effects on different types of waves might be incomplete. One explanation proposed within the standard model is that the light was emitted later by explosive jets rather than simultaneously with gravitational waves. While plausible within known astrophysical processes, this interpretation is invoked after the fact to reconcile an otherwise unexpected discrepancy.

The EAC provides a more natural explanation by proposing that gravity and light propagate through distinct physical mechanisms—gravity traveling through the curvature of spacetime itself, while light is influenced by a quantum-based refractive medium, introducing a time delay. The "halo of gravity" at the edge of the observable universe, as proposed in the EAC, can delay the arrival of light while leaving gravitational waves unaffected.

By resolving this discrepancy, the EAC offers a compelling alternative to the inflationary Big Bang model. This highlights the need to reconsider how gravitational and electromagnetic waves interact with cosmic structure—an essential factor in refining our understanding of the universe.

The Path Forward

By recognizing where the EAC converges with the current cosmology, we can focus research efforts on its unique contributions while building upon established successes. This approach invites collaboration rather than confrontation, suggesting specific areas for investigation:

- Precise predictions for gravitational lensing effects

- Observable signatures that could distinguish between initial condition mechanisms

Like the unification of electricity and magnetism into electromagnetism, perhaps these seemingly different cosmological perspectives will ultimately reveal deeper underlying principles about the nature of space, time, and observation itself.

A Bridge to Understanding

This convergence reminds us that scientific progress often comes not from wholesale replacement of existing theories, but through careful refinement and reexamination of fundamental assumptions. The EAC offers new insights into cosmic origins and the nature of observation while preserving the hard-won successes of modern cosmology.

As we stand at this theoretical bridge between models, we glimpse the possibility of a deeper understanding–one that honors both the established wisdom of standard cosmology and the novel insights of the Expanding Awareness framework. Like the universe itself, our comprehension expands not through rejection of what came before, but through the discovery of new perspectives that encompass and transcend existing knowledge.

This meeting of models invites us to explore the cosmos with renewed wonder, armed with complementary tools for understanding its deepest mysteries. The journey forward promises not just refined calculations and more precise predictions, but a richer, more complete picture of our universal home.

The Big Block

Rather than a singularity, the Expanding Awareness Cosmology posits a "Big Block" as the universe's precondition state. Imagine a vast, crystalline structure of pure potential, composed of tightly packed Higgs bosons in both positive and negative states. While speculative, this concept provides a simple and natural explanation for several observed cosmic features, including the matter-antimatter asymmetry and the cosmic microwave background radiation.

Unlike the hypothetical inflation field, which appears and vanishes as needed, or the notion of repulsive gravity, the Big Block represents a fundamental state of the universe—existing before the onset of time and change. In this state, the Higgs field is not yet active as a field of excitations but instead exists as a unified, unchanging entity.

Although the term "Big Block" is metaphorical, it conveys the essence of this pre-existing state: a dense, compact configuration of the Higgs field, capable of setting the initial conditions for the universe and driving the emergence of matter and energy.

This concept offers an alternative to the singularity in the Big Bang model. Instead of emerging from an infinitely dense and hot point, the universe could have originated from a more stable and unified state—the Big Block. By challenging traditional assumptions about cosmic origins, this framework opens new avenues for exploring the fundamental nature of reality.

The Abundance of Light Elements

The observed abundance of light elements in the universe aligns seamlessly with the Expanding Awareness Cosmology. The cooling period following the "Let there be light" event—the initial matter-antimatter annihilation—created the necessary conditions for nucleosynthesis. The Big Bang model attributes this cooling to the expansion of space.

In contrast, the Expanding Awareness Cosmology suggests that cooling occurred because most of the energy had already been released, with only isolated pockets of antimatter remaining. As gravity pulled these pockets toward matter, minor annihilation events followed the initial blast, contributing to localized variations in temperature.

Like a cosmic chef adjusting temperatures in different parts of an oven, these temperature variations during nucleosynthesis could have influenced neutron capture rates, creating regions where elements formed at different rates. This effect might explain why we observe less lithium-7 than standard models predict—a discrepancy known as the "lithium problem."

Just as uneven heating alters the outcome of a chemical reaction, slight differences in cosmic temperature could have affected element formation.

These temperature pockets may have also led to subtle variations in helium and deuterium abundances. By studying light from distant quasars, researchers could detect these variations much like geologists analyze ancient minerals to understand Earth's early conditions. As the universe transitioned from its initial high-energy state to a lower-energy state, the precise temperatures and densities required for nuclear fusion emerged naturally.

The observed ratios of hydrogen, helium, and other light elements align with Big Bang nucleosynthesis but within the specific conditions of the Expanding Awareness Cosmology. The "Big Block" of tightly packed Higgs bosons, with its inherent energy density, provided the raw material for these elements to form. As the universe cooled and particles interacted, the fundamental forces of nature guided the formation of these elemental building blocks.

The abundance of light elements serves as a chemical fingerprint of the universe's earliest moments. Rather than being solely a relic of the Big Bang, it emerges as a direct consequence of the Big Block and the cooling period that followed. This evidence further supports the Expanding Awareness Cosmology, demonstrating its ability to account for key cosmological observations without relying on the assumption of an expanding universe.

Incorporating the Cosmic Microwave Background

Within Expanding Awareness Cosmology, the Cosmic Microwave Background (CMB) radiation emerges naturally from the universe's initial conditions. The Big Block, with its inherent uniformity, provides an ideal foundation for understanding the CMB's observed characteristics. The remarkable uniformity of the CMB across the sky mirrors the uniformity of the Big Block, suggesting a deep connection between the universe's earliest state and the radiation we observe today. Like a cosmic echo of the universe's first moments, the CMB carries the imprint of the Big Block across vast stretches of space and time.

Small temperature fluctuations in the CMB originate from two primary sources: quantum fluctuations within the Big Block and delayed annihilation events, where residual antimatter, drawn toward matter by gravity, undergoes secondary interactions.

Like ripples spreading through a vast cosmic pond, these fluctuations introduce subtle variations in the density and distribution of energy and matter. As the universe cools from its initial high-energy state, these variations become permanently encoded in the CMB as temperature fluctuations.

The CMB's blackbody radiation profile and temperature evolution reflect the cooling phase that followed the initial matter-antimatter event. As the universe transitioned from the high-energy conditions of the Big Block to a lower-energy state, the CMB cooled in response.

Imagine the early universe as a vast ocean of light and matter. Standard theory interprets the CMB as waves breaking uniformly on a cosmic shore. However, just as underwater hot springs create localized warm spots in an ocean, late-time particle annihilation could have produced pockets of reheated gas. These thermal eddies may have subtly rippled through spacetime, leaving imprints in the CMB's otherwise perfect blackbody spectrum—delicate distortions that next-generation satellites may finally detect, much like sensitive instruments can measure the tiniest temperature variations in Earth's oceans.

In this framework, the CMB acts as a bridge between the universe's initial conditions and the large-scale structures we see today. It is not merely a relic of the Big Bang but a direct consequence of the Big Block and the evolving processes that shape our cosmic reality.

Scientific Genesis

The ancient words of Genesis take on new meaning in this context: *"In the beginning... darkness was upon the face of the deep."* Before the first moment, the universe existed in perfect stasis—no motion, no change, no time. Then came the first event: *"Let there be light."* In modern terms, this marks the instant when positive and negative Higgs bosons first became aware of each other, triggering the matter-antimatter annihilation that defined the first Planck moment of time itself.

From this primordial state emerged a cascade of transformative events. As particles formed and mass manifested through interactions with the Higgs field, space itself became foreshortened by gravity. Like ripples spreading across a pond, the effect of gravity propagated outward at the speed of light as particles began to sense and interact with their surroundings.

This concept demands a profound shift in our understanding. The universe cannot be conceived as a unified system where every particle instantaneously interacts with every other particle. Instead, each particle exists within its own expanding sphere of awareness, gradually discovering and engaging with its cosmic neighbors. Like a social network growing one connection at a time, the universe builds its complexity through an ever-expanding web of relationships.

This framework naturally gives rise to Einstein's long-sought cosmological constant through an elegant balance of forces. The mathematics emerges from simple geometric principles in three-dimensional space, expressed through a fundamental equation:

$$\frac{GM(r)m}{r^2} + \frac{GM(t)m}{ct^2} = 0$$

This equation captures a profound balance in the universe. The first term G represents the classical Newtonian gravitational force pulling matter inward, while the second term represents a counterbalancing force that increases with the total mass of the observable universe. Like a perfectly balanced scale, these forces achieve equilibrium at every point.

The internal force follows the familiar inverse square law of gravity, where $M(r)$ represents the mass enclosed within radius r. But the external term introduces something remarkable: a force that depends on both the total mass $M(t)$ and the age of the universe t, modulated by the speed of light c. This isn't an arbitrary mathematical construct but emerges naturally from the finite propagation speed of gravitational effects.

The density ρ_0 remains constant as our sphere of awareness expands, ensuring that $M(t)$ increases proportionally with the universe's observable volume. This creates a dynamic equilibrium—not between mysterious dark energy and gravitational collapse, but between our expanding awareness of distant matter and the accumulated gravitational effects we can measure.

Like a masterful architectural dome that channels forces into perfect stability, this equation reveals how the universe maintains its structure through precise geometric balance. The cosmological constant emerges not as an ad hoc addition but as a natural consequence of gravity's finite propagation speed in a static universe. Think of it like the growth of human knowledge: each discovery reveals new questions, creating an illusion of endless expansion, yet this growth occurs within the bounded reality of what exists to be known. Similarly, our universe appears to expand as more distant regions come into causal contact, while maintaining its fundamental static nature.

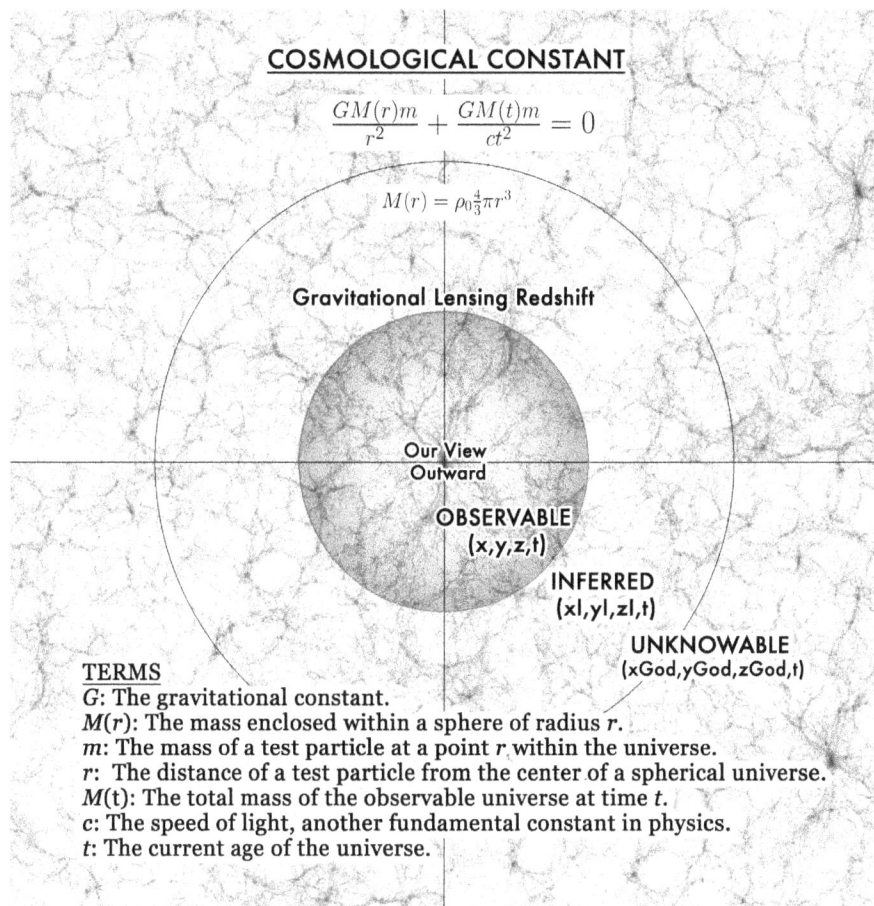

COSMOLOGICAL CONSTANT

$$\frac{GM(r)m}{r^2} + \frac{GM(t)m}{ct^2} = 0$$

$$M(r) = \rho_0 \tfrac{4}{3}\pi r^3$$

Gravitational Lensing Redshift

Our View
Outward

OBSERVABLE
(x,y,z,t)

INFERRED
(xI,yI,zI,t)

UNKNOWABLE
(xGod,yGod,zGod,t)

TERMS
G: The gravitational constant.
$M(r)$: The mass enclosed within a sphere of radius r.
m: The mass of a test particle at a point r within the universe.
r: The distance of a test particle from the center of a spherical universe.
$M(t)$: The total mass of the observable universe at time t.
c: The speed of light, another fundamental constant in physics.
t: The current age of the universe.

This model resolves one of Einstein's most profound insights—one that science has struggled to fully appreciate. His special relativity placed the observer at the center of all physical understanding, yet we've often tried to maintain the illusion of an objective, frame-independent perspective. The expanding awareness cosmology embraces this observer-centrality as fundamental to nature itself.

Had Einstein extended his famous light-beam thought experiment to consider an observer at two distant points looking back, he might have discovered this framework himself. The cosmological constant he sought might have emerged naturally, providing an alternative explanation for the redshift Hubble would later observe. Given Einstein's experience with gravitational lensing in the Eddington experiment, he was close to unifying these concepts.

The Edge Observer Thought Experiment

To understand how the universe can appear to expand while maintaining an underlying structural continuity, we must extend Einstein's thought experiments on light and observation to their logical conclusion. Consider a thought experiment involving two observers separated by vast cosmic distances–ourselves and a hypothetical observer we'll call Luke, positioned in a distant galaxy far, far away.

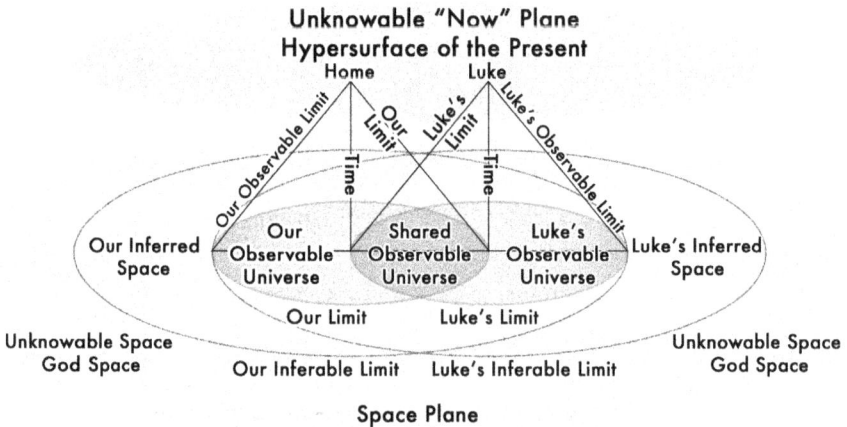

When we look toward Luke's location, we see what appears to be a young galaxy, its light having traveled billions of years to reach us. This image does not represent Luke's present reality but an ancient snapshot of his cosmic neighborhood. Likewise, when Luke observes our galaxy, he sees it as it was billions of years ago—long before our current moment.

This symmetry reveals something fundamental about the nature of space and time. Between our two positions lies the same physical universe—the same matter, the same energy, the same fundamental reality—viewed from two different points in time. Like two artists painting the same landscape from opposite hillsides, but at different historical moments, we perceive the same cosmos through distinct temporal lenses.

A key insight emerges when we consider that both observers exist in their own "Now"—the moment in which all available light and gravitational signals have converged to shape their present reality. This "Now" is not just a point in time but the sum of all past observable history at a given location in space. Often described as the "hyperspace of the present," this universal Now is uniquely defined by each observer's vantage point. The Unknowable Now Plane represents the totality of simultaneous events across the cosmos—an objective reality that remains hidden from any single observer, accessible only to an all-encompassing intelligence, if such an entity exists.

This realization transforms our understanding of cosmic structure. At every point in space, information propagates at the speed of light through both electromagnetic radiation and gravitational effects. What we perceive as a vast, evolving cosmos is in reality a web of interwoven pasts, each observer limited to a personal horizon of information. The universe, rather than expanding into the unknown, unfolds as an ever-growing tapestry of observed histories, shaped by the fundamental limits of light and gravitational propagation.

Video Games and Simulation

You are an explorer, a cosmic detective uncovering the deepest mysteries of existence. Your mind—an extraordinary instrument of discovery—transforms abstract concepts into navigable landscapes of insight, allowing you to see beyond the limits of direct observation.

Just as Newton saw the laws of gravity in a falling apple and Einstein imagined riding a beam of light to explore relativity, you too can unlock profound cosmological insights—through an unexpected lens: the intricate worlds of strategic exploration in 4X video games.

Imagine yourself as a player in these virtual universes. You begin with limited awareness, a tiny spark of potential surrounded by unknown terrain. With each move, each interaction, you gradually unveil a pre-existing world—a process that mirrors our scientific journey of cosmic understanding. This isn't mere entertainment; it's a powerful metaphor for human discovery.

Consider the parallels: In a 4X game, the complete map exists before you begin playing, but you can only access regions you've explored. Similarly, our universe may not be expanding into nothingness, as traditional models suggest. Instead, you are expanding your sphere of awareness into a pre-existing cosmic space, methodically revealing what has always been there.

Your scientific instruments function like a game's "fog of war"—gradually revealing hidden cosmic landscapes, turning the unknown into the known. Each observation is a triumph, each discovery a new tile added to humanity's expansive map of reality.

The Expanding Awareness Cosmology isn't just a theoretical construct—it's an invitation. An invitation for you to see yourself as an active participant in universal understanding, not a passive observer. The fundamental laws of physics aren't obstacles; they are the rule set waiting for your intellect to decode, to master, to transform.

No alien civilization handed us an owner's manual for consciousness. But in its absence, you have something far more powerful: curiosity, reasoning, and an insatiable drive to understand. Your journey of discovery is the manual—written in real-time, with each breakthrough, each moment of insight.

The mathematics of information propagation and the geometry of causal horizons aren't dry academic concepts. They are your tools, your keys to unlocking deeper comprehension. Like a skilled game designer crafting intricate rule systems, the universe offers you a complex, beautiful simulation to explore.

You are not just watching the game—you are playing it. With every observation, every theory tested, and every horizon expanded, you advance humanity's greatest quest: the pursuit of cosmic understanding.

Are you ready to explore?

Observable, Inferable, and Unknowable

Our Edge Observer thought experiment reveals a profound truth about how we come to know the universe. It's like peeling back layers of an onion, where each layer represents a different domain of knowledge:

The Observable Universe

The innermost layer is the realm of direct experience—everything we can see, measure, and detect with our instruments. This is the domain of scientific certainty, bounded by the speed of light and the finite age of the universe. Within this sphere, we can map the distribution of matter, detect gravitational waves, and analyze the light from distant stars.

The Inferable Universe

Beyond the limits of direct observation lies a realm accessible through logical deduction and extrapolation. Like a detective piecing together clues, we can infer the existence of things we cannot directly see. The uniformity of the observable universe suggests that similar structures exist beyond our current horizon. Likewise, we reconstruct past events—such as the formation of the first stars—by analyzing the evidence available in the present.

The Unknowable

Even with our most advanced instruments and deepest reasoning, there remains a realm forever beyond our grasp. This outermost layer is the domain of the truly unknowable. It includes questions about the ultimate origins of the universe, the nature of consciousness, and the full scope of existence. These mysteries are not meaningless, but they exist beyond the reach of scientific investigation, leaving us to explore them through philosophy, speculation, and imagination.

Inferring on the Unknowable

While the unknowable remains forever outside our direct experience, we can still make inferences about it. Like an artist sketching the outline of a hidden figure, we can use logic, existing knowledge, and explanatory power to make educated guesses about what lies beyond the veil.

The "Big Block": To explain the observed universe, we infer the existence of an initial state—a "Big Block" of tightly packed Higgs bosons. This "Big Block" is not something we can observe, but its existence offers a plausible explanation for the universe's structure and evolution. Crucially, we posit that this "Big Block," while perfectly symmetrical in its overall structure, contained a slight initial asymmetry in the distribution of positive and negative Higgs bosons. This inherent asymmetry, though its origin lies beyond our current understanding, serves as the seed for the matter-antimatter imbalance we observe in the universe today.

Initial Distribution of Matter: We assume that all matter existed from the beginning in its current location. This is a "brute fact" style assumption, akin to accepting the values of fundamental constants without a deeper explanation.

"Halo of Gravity": To explain the redshift of distant galaxies, we infer the existence of a "halo of gravity" at the edge of the observable universe. This halo, generated by the cumulative gravitational influence of unseen matter, acts as a cosmic lens, distorting the light that reaches us.

It's important to remember that these inferences are not arbitrary guesses. They are grounded in logic, consistency with existing knowledge, and their ability to explain observed phenomena.

Comparison with the Big Bang

It's important to recognize that the Big Bang theory, like any cosmological model, relies on inferences about the unknowable. For example:

Origin of Space

The Big Bang theory assumes that space itself came into existence at the moment of the Big Bang. This raises a fundamental question: What, if anything, existed before space? Since space and time are intertwined, the idea of a "before" may not even be meaningful within this framework, yet the assumption remains beyond empirical validation.

Cosmic Inflation

To explain certain observations, the theory of cosmic inflation posits an early period of extremely rapid expansion. However, key aspects of this model remain unresolved:

- What was the source of the energy driving inflation?

- What mechanism initiated and ended inflation?

- What is the nature of the inflation field?

These questions remain completely untestable, making inflation an inferred construct rather than an empirically established process.

While inflation elegantly accounts for cosmic flatness, it may be an unnecessary complication. The universe's flatness could simply be an intrinsic property, much like the speed of light is accepted as a fundamental constant without requiring an underlying mechanism. Sometimes, Occam's razor suggests that instead of constructing elaborate origin stories, we should consider whether the observed features of the universe are simply fundamental aspects of reality.

Science, Religion, and the Unknowable

This division of cosmic knowledge clarifies the relationship between science and religion. Science, grounded in observation and inference, excels at exploring the observable and inferable realms. However, it reaches its limits at the boundary of the unknowable—questions beyond empirical investigation.

This is where religion and philosophy step in. They engage with the mysteries that science cannot resolve, offering frameworks for meaning, purpose, and the ultimate nature of reality. Yet, science also attempts to push the bounds of the knowable and occasionally crosses the line into the realm of religion.

When scientific models speculate beyond testable predictions—such as proposing unobservable multiverses or self-creating universes—they enter a space traditionally occupied by philosophy and metaphysics. Rather than being in conflict, these disciplines complement one another, each addressing different aspects of human understanding.

By acknowledging the limits of scientific inquiry and the existence of the unknowable, we adopt a more holistic approach to understanding the universe. This perspective embraces both the rigor of empirical investigation and the insight of philosophical and spiritual reflection.

It allows us to explore the cosmos with both precision and wonder, recognizing that while some mysteries may be solved, others may forever remain beyond our reach.

Mathematical Bridges Across Three Domains

Just as complex numbers revolutionized our understanding of mathematical spaces, we introduce a framework that explicitly maps the transitions between observable, inferable, and unknowable domains. This framework employs two operators: 'I' for inferred space and 'God' for the absolute boundary of knowability.

The choice of the full word "*God*" as an operator serves a specific purpose. Mathematical exploration often risks drifting into unverifiable abstractions, and without clear boundaries, researchers may unknowingly move beyond testable science, as seen in aspects of modern string theory. This notation acts as a conceptual safeguard:

- When an idea moves into the inferred domain ('I'), it raises a yellow flag, signaling that conclusions rely on extrapolation rather than direct observation.

- When analysis crosses into the '*God*' realm, the red flag waves vigorously, marking a transition into the truly unknowable—where scientific certainty and credibility become precarious.

This structured approach ensures that future explorations remain grounded, preventing the unintentional drift from rigorous inference into speculative metaphysics.

The Three-Domain Framework:

- Observable space: Described by traditional coordinates (x,y,z,t)

- Inferred space: Described by coordinates with the I operator

- Unknowable domain: Marked by the *God* operator

Consider a point P in space:

- In observable space: $P(x,y,z,t)$

- In inferred space: $P(xI,yI,zI,t)$

- At the boundary of inference: $P(xGod,yGod,zGod,t)$

Like the complex plane's perpendicular axes representing real and imaginary components, our universe possesses three distinct but connected domains. The God operator serves a dual purpose:

- Mathematical: It marks the absolute boundary beyond which inference fails

- Philosophical: It provides a rigorous reminder of science's inherent limits

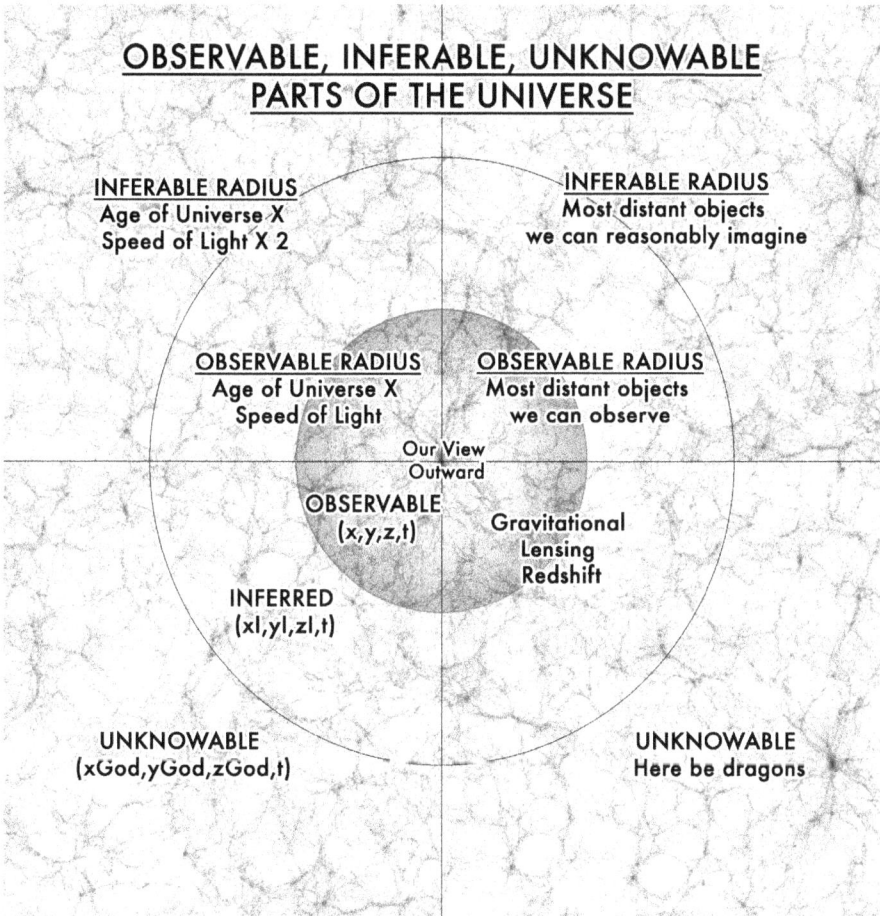

OBSERVABLE, INFERABLE, UNKNOWABLE PARTS OF THE UNIVERSE

INFERABLE RADIUS
Age of Universe X
Speed of Light X 2

INFERABLE RADIUS
Most distant objects
we can reasonably imagine

OBSERVABLE RADIUS
Age of Universe X
Speed of Light

OBSERVABLE RADIUS
Most distant objects
we can observe

Our View
Outward

OBSERVABLE
(x,y,z,t)

Gravitational
Lensing
Redshift

INFERRED
(xl,yl,zl,t)

UNKNOWABLE
(xGod,yGod,zGod,t)

UNKNOWABLE
Here be dragons

Critical Properties:

- Just as $i^2 = -1$ defines the behavior of imaginary numbers, God terms in equations become mathematically undefined or indeterminate

- Any equation containing *God* terms in intermediate steps is immediately flagged as overstepping scientific bounds

- The *God* operator identifies when theoretical frameworks inadvertently make claims about the unknowable

If r represents the radius of our observable universe (approximately 13.8 billion light years), then points at radius R can be expressed as:

$$R = r + kI + mGod$$

where:

- r represents the directly observable component

- k represents the additional distance into inferred space

- m marks the transition to the unknowable

This framework provides both mathematical precision and philosophical clarity about the boundaries of scientific inquiry. Like ancient cartographers writing "Here Be Dragons" at the edge of known waters, but with mathematical rigor, we acknowledge the boundary between the knowable and unknowable with unmistakable clarity.

The Bridge to Universal Balance

The Edge Observer thought experiment reveals how the cosmological constant emerges naturally from fundamental principles of causality and the finite speed of gravity. Like ripples spreading from two stones dropped into a pond, gravitational influence expands spherically from every point in space, creating overlapping domains of interaction that precisely balance each other.

Consider the equation:

$$\frac{GM(r)m}{r^2} + \frac{GM(t)m}{ct^2} = 0$$

The first term represents what each observer can "see" gravitationally within their immediate sphere—like the clear water directly around each dropped stone. Just as Luke and we observe different historical snapshots of each other's galaxies, we each experience a unique gravitational sphere defined by our local mass distribution $M(r)$.

The second term captures something more subtle: the accumulated gravitational effect of all matter that exists beyond our immediate observational sphere, $M(t)$. Like the complex interference patterns that develop when the ripples from our two stones meet and interact, this term represents the sum of gravitational influences still propagating through space from distant masses we haven't yet "seen."

The temporal factor ct^2 in the denominator isn't merely mathematical convenience—it represents the fundamental limitation imposed by gravity's speed-of-light propagation. Just as Luke and we can only see each other's past, we can only feel gravitational effects from masses whose influence has had time to reach us.

This creates a perfect balance. As our sphere of gravitational awareness expands at light speed, we discover new masses that exert their influence on us. However, these same masses, through the geometry of spherical propagation, contribute to the counterbalancing external force that prevents universal collapse. Like a cosmic ballet choreographed by the laws of physics themselves, each new discovery maintains the eternal dance of forces in perfect equilibrium.

While the balance of gravitational interactions emerges naturally in the expanding causal structure of the universe, Einstein's Field Equations (EFE) assume a continuous, smoothly connected spacetime fabric. However, to fully account for the **finite speed of gravity** and the **limitations imposed by causal horizons**, modifications to EFE may be necessary. These adjustments would ensure that gravitational influence is correctly limited to **causally connected regions** while preserving observational consistency.

Modifying Einstein's Field Equations

To modify Einstein's Field Equations (EFE) to explicitly incorporate the finite propagation speed of gravity and ensure that only particles within the causal horizon (defined as the age of the universe multiplied by the speed of light) remain in causal contact, several key adjustments must be considered.

1. Incorporating a Finite Propagation Speed for Gravity

General relativity already implies that gravitational interactions propagate at the speed of light, as gravitational waves arise naturally from the linearized EFE. However, standard formulations often treat gravity as an instantaneous field when dealing with static or weak-field approximations. To explicitly enforce finite propagation, the following modifications could be introduced:

- Retarded Potential Formalism: Replace instantaneous interactions with terms that depend on the past positions and states of sources, ensuring that gravitational effects reflect finite propagation delays rather than acting instantaneously.

- Integration Over the Past Light Cone: Modify gravitational influence calculations to integrate only over causal regions, ensuring that no information propagates faster than light.

2. Limiting Causal Interactions to Within the Horizon

To ensure that gravitational interactions are restricted to causally connected regions:

- Introduce a Causal Cutoff in the Stress-Energy Tensor

- The stress-energy tensor $T^{\mu\nu}$, which serves as the source term in EFE, could be modified to exclude contributions from beyond the causal horizon.

- This could be implemented mathematically using a Heaviside step function or another cutoff mechanism applied to $T^{\mu\nu}$, ensuring that only matter and energy within a distance $d = c \times t_{universe}$ contribute to gravitational effects.

3. Modifying the Einstein-Hilbert Action

To incorporate causal constraints into the fundamental formulation of general relativity, the Einstein-Hilbert action could be modified:

$$S = \int [R + \mathcal{L}_{\text{matter}}] \sqrt{-g}\, d^4x$$

where R is the Ricci scalar curvature. To reflect finite propagation speeds and causal limits, modifications could include:

Adding Terms Dependent on Retarded Potentials

- Integrating over past light cones to ensure that gravitational interactions do not exceed the speed of light.

Introducing Functional Dependencies on R or $T^{\mu\nu}$ That Enforce Causality

- Explicitly incorporating causal constraints within the field equations rather than relying on external conditions.

4. Gravitational Radiation and Causality

While gravitational waves already propagate at c, additional modifications might be required to ensure causality in dynamic spacetimes, particularly in extreme environments such as:

Binary Mergers

- Incorporating causal constraints in rapidly evolving gravitational fields.

Cosmological Models with Evolving Curvature

- Ensuring that changes in the fabric of spacetime itself remain fully consistent with causal limits.

Challenges and Implications

Modifications to general relativity must account for several critical challenges:

Stable Orbits:

- Any changes must preserve stable orbital dynamics. Some models introducing finite propagation speeds have led to instabilities or tangential accelerations.

Energy Conservation:

- The revised formulation must still satisfy local conservation laws.

Observational Consistency:

- Any modifications must remain consistent with precise measurements, such as gravitational wave detections (e.g., GW170817), which confirm that gravity propagates at c.

While Einstein's original formulation already respects causality, making this principle explicit in the equations would involve incorporating retarded potentials and causal cutoffs into both the field equations and their source terms. These refinements would ensure that gravitational effects remain confined to regions within causal contact, preserving both mathematical consistency and observational accuracy.

A Universe of Awakening Awareness

The Edge Observer thought experiment reveals another profound insight: the universe does not consist of a single expanding bubble of awareness, but rather countless overlapping spheres of causal contact, each expanding at the speed of light. Like morning fog lifting to reveal a landscape dotted with dewdrops, each reflecting its own view of the world, every particle begins its existence in isolation, gradually expanding its awareness as it interacts with its cosmic neighbors.

Discovering a Shared Reality

Let's return to Luke, our distant observer from the Edge Observer thought experiment, and examine the transitional points of increasing awareness between us.

At first, we share no common data with Luke. No part of the universe is observable to both parties. Our awareness is completely separate. But as time passes and the universe expands, the halfway point between us becomes visible to both, marking the first moment of shared awareness. At this point, we each see the same collection of matter and energy—though we still cannot yet observe each other's primordial beginnings.

The next major transition is the starting point of the Edge Observer thought experiment itself: the moment when both observers can finally see the other's cosmic origins. We see his beginnings, and he sees ours. At this stage, we can also infer regions of space that Luke has already observed but remain beyond our current reach, expanding our indirect awareness of the cosmos.

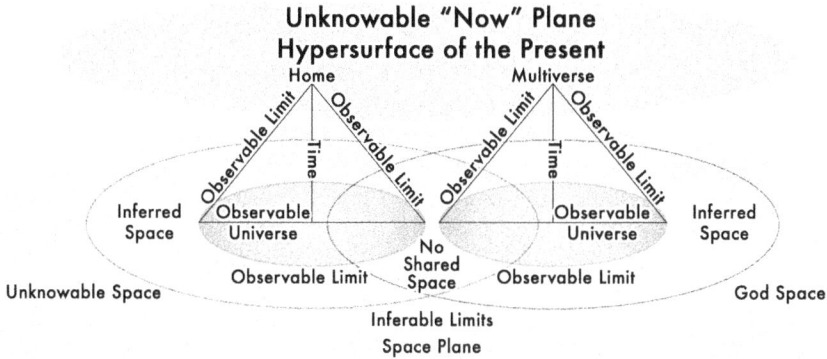

The Initial Growth of Countless Realities

This perspective reshapes our understanding of early cosmic structure formation. Consider a region of particularly high density in the aftermath of matter-antimatter annihilation. In conventional models, such regions would rapidly form black holes, but their growth would be constrained by radiation pressure, which pushes surrounding matter away. However, this view overlooks the role of expanding causal awareness in the early universe.

In these primordial moments, countless black holes would form simultaneously, each within its own isolated bubble of causal contact. Like separate kingdoms unaware of their neighbors, these regions would evolve independently until their spheres of gravitational influence began to overlap. When they did, the consequences would be dramatic.

Imagine a billion solar masses of matter in a localized area, initially forming many small, separate black holes. As their gravitational awareness expands and intersects, they merge rapidly, coalescing into a single supermassive structure. In this way, the universe would quickly populate itself with primordial black holes, jump-starting the formation of the cosmic web.

This process of merging causal spheres offers a compelling solution to one of modern cosmology's greatest mysteries—the existence of supermassive black holes (SMBHs) in the early universe. Instead of requiring billions of years for gradual accumulation, these colossal objects could form rapidly, as multiple independent regions come into causal contact. Like streams converging into rivers, which then flow into mighty deltas, cosmic structure formation follows the expanding boundaries of gravitational awareness. Each particle, clump of matter, and nascent black hole exists initially in its own light-speed-expanding bubble of causality, merging and combining as these bubbles intersect across space and time.

Implications for Early Universe Observations

These causality-based refinements to Einstein's Field Equations (EFE) are not just theoretical adjustments; they provide a framework for addressing some of the most perplexing observations in early universe cosmology.

One of the biggest unresolved mysteries is the existence of supermassive black holes (SMBHs) and fully mature galaxies at high redshifts ($z > 7$)—objects that appear too massive and too well-formed to have developed within the standard timeline of structure formation. If gravitational influence is constrained by finite-speed causal horizons, then early black hole mergers and galaxy evolution could have occurred much more rapidly than previously expected.

Supermassive Black Hole Formation

Instead of relying on gradual accretion or exotic mechanisms like direct-collapse black holes, these objects may have assembled rapidly through gravitational mergers, as expanding awareness spheres overlapped. This could resolve the apparent paradox of SMBHs forming too quickly for standard ΛCDM (Lambda Cold Dark Matter) models.

Fully Formed Galaxies at High Redshift

Observations of massive, well-structured galaxies within the first few hundred million years of the Big Bang present a major challenge to ΛCDM models. If causal constraints affect how early mass concentrations interact gravitationally, it could explain how such galaxies formed and stabilized far more efficiently than standard models predict. By incorporating causal restrictions into our models of gravitational interaction, we may uncover a missing piece of early structure formation—one that resolves long-standing tensions between observation and theory.

But the implications extend beyond early supermassive black holes. This framework provides a new way to understand the entire spectrum of cosmic structures—from diffuse galaxies to massive gravitational lenses. To fully grasp these consequences, we must now explore how this model redefines our interpretation of cosmic time and space.

Gravity and Quantum Reality

The connection between gravity and the quantum world has long been physics' most stubborn puzzle. Reconciling quantum mechanics with general relativity is like trying to translate between two alien languages, each evolved on entirely different foundations. Yet, the Higgs boson, that most elusive of particles, may provide the bridge we have been seeking.

Consider the extraordinary effort required to observe a single Higgs boson: a 17-mile circular accelerator, generating unimaginable energies, all to glimpse a particle that exists for less than a trillionth of a second. This extreme elusiveness is itself a profound clue. Like archaeological remnants of an ancient civilization, the scarcity and fragility of Higgs bosons may be evidence of their primordial role in universe formation.

The Higgs Field as the Fabric of Space

In this framework, the Higgs field is not merely the provider of mass—it is the substrate of space itself. The pre-existing "Big Block" of densely packed positive and negative Higgs bosons explains both their current rarity and their connection to gravity through an elegant mechanism: the foreshortening of space by information.

Imagine a pristine void, untouched by mass or energy—perfect, undisturbed space. Now, introduce information into this space, perhaps in the form of quantum vibrations, similar to those postulated by string theory. These vibrations create a subtle but critical effect: they foreshorten space itself. Like comparing the length of a taut string to one that vibrates in waves, the presence of quantum information shortens the effective distance between points, altering the very structure of space.

A New Language for Gravity

This foreshortening effect provides an entirely new way to describe gravity—one that matches Einstein's curved spacetime while emerging naturally from quantum principles. When massive objects interact with the Higgs field, they create patterns of information that foreshorten space around them. Other objects, moving along what they perceive as straight lines through this altered space, follow exactly the curved trajectories we observe as gravitational motion.

This framework eliminates the need for a graviton particle, resolving one of quantum gravity's most persistent challenges. Instead of requiring a force-carrying particle that has stubbornly resisted detection, gravity emerges naturally as a consequence of how quantum information carried by the Higgs field alters the texture of space itself.

Implications for Unifying Gravity and Quantum Mechanics

The implications ripple far beyond theoretical elegance. This mechanism provides a direct bridge between quantum mechanics and classical gravity, explaining why:

- Gravitational effects manifest at all scales, while

- Quantum effects remain confined to microscopic interactions.

Like a fractal pattern, revealing different structures at different scales, the same underlying principle—information-based foreshortening—manifests differently depending on our observational perspective. This model suggests that instead of being fundamentally incompatible, gravity and quantum mechanics are two expressions of the same deep, information-driven structure of reality—one that emerges naturally from the interaction of space, mass, and information itself.

Space, Gravity, and Light

Across the vast expanse of the cosmos, two fundamental messengers carry information through space: gravity and light. Though they travel at the same speed, they convey reality in profoundly different ways. Understanding this distinction not only clarifies one of the key challenges to the Expanding Awareness framework but also reveals deeper insights into the structure of the universe.

Two Messengers, Two Mediums

Imagine a stone dropped into a perfectly still pond. The impact creates two distinct effects: ripples spreading across the water's surface and light reflecting off the disturbed waves. Both carry information about the event, yet they interact with their surroundings in entirely different ways.

The ripples move through the water itself, shaped by the medium's geometry, while the light bounces off each wavelet, creating a complex interplay of reflection and refraction.

A similar distinction exists between gravitational waves and electromagnetic radiation as they traverse space:

Gravitational Waves

- Propagate through the fabric of space itself

- Follow pure geodesics determined by spacetime geometry

- Experience minimal interaction with matter

- Maintain their fundamental character across cosmic distances

Electromagnetic Waves

- Travel through space while interacting with quantum fields

- Subject to scattering, absorption, and reemission

- Experience cumulative lensing effects

- Path can be distorted by intervening matter and energy

This distinction is essential for interpreting cosmic observations. When both gravitational waves and light are detected from the same celestial event, such as the merger of two neutron stars, the gravitational signal typically arrives first. It follows the most direct path through spacetime, largely unaffected by intervening matter. The light, however, arrives slightly later, its journey shaped by interactions with dust, gas, and gravitational fields, each introducing delays and distortions.

Observations of split supernovas, where light from a single explosion is gravitationally lensed by a massive intervening galaxy, present an intriguing thought experiment. In such cases, the light can take multiple paths around the lens, arriving at different times and at different locations in the sky—sometimes years apart. However, gravitational waves from the same event would behave differently. Since they travel the shortest possible geodesic path through spacetime, they would be detected only once. This contrast underscores the fundamental difference in how gravity and light interact with the universe, reinforcing the role of gravitational waves as pure geometric signals, unaffected by the same distortions that complicate the journey of light.

By recognizing how gravity and light travel through space in distinct ways, we gain a clearer understanding of the universe's fundamental structure and the nature of the signals we receive from its most energetic events.

Redshift and Cosmic Structure

This differential propagation illuminates a key aspect of the Expanding Awareness cosmology. The observed redshift of distant galaxies, traditionally interpreted as evidence for universal expansion, takes on new meaning when we consider the cumulative effects of gravitational lensing on light.

The "halo of gravity" at the boundary of our observable universe affects electromagnetic radiation profoundly, while gravitational waves pass through relatively undisturbed. Like sound waves traveling through layers of different density in the ocean, light experiences progressive distortion as it traverses regions of varying gravitational potential.

The gravitational waves, analogous to pressure waves moving through the water's bulk, maintain their fundamental character. This explains why gravitational wave observations often provide cleaner, more direct measurements of cosmic distances.

This observed discrepancy in arrival times poses a direct challenge to the Big Bang model, which predicts that gravitational waves and light from the same source should arrive simultaneously due to their equal stretching by cosmic expansion. The EAC, on the other hand, naturally accounts for this discrepancy through its concept of differential propagation and the lensing effect of the 'halo of gravity.

Mathematical Framework

The distinct ways in which gravitational waves and electromagnetic waves propagate through space reveal key differences in how they interact with the universe.

Gravitational waves move along the natural curves of spacetime, following the shortest possible path determined solely by spacetime geometry. They travel unhindered through cosmic structures, largely unaffected by matter or energy along the way.

Electromagnetic waves, on the other hand, must interact with charged particles and fields as they propagate. Their journey is shaped by scattering, absorption, and lensing effects, altering both their trajectory and timing. These additional interactions introduce variations that do not affect gravitational waves in the same way.

Observational Consequences

This framework makes several testable predictions:

1. Arrival Time Patterns

- Gravitational waves from cosmic events should consistently arrive before their electromagnetic counterparts.

- The delay between the two signals should vary depending on the distribution of intervening matter, which affects light but not gravitational waves.

2. Clean Distance Measurements

- Gravitational wave signals should provide more accurate distance measurements than light-based methods, as they are not distorted by gravitational lensing.

- Comparing gravitational wave distances with electromagnetic distances could offer a new way to measure the effects of cosmic structure on light.

3. Structure Formation Signatures

- The ability of gravitational waves to travel undisturbed may help explain how early cosmic structures formed so rapidly.

- Detecting ancient gravitational waves could provide a new historical record of galaxy formation, offering insights that electromagnetic signals alone cannot reveal.

Early Universe Implications

The difference in how gravitational and electromagnetic waves propagate is especially important for understanding the early universe. Unlike light, which is scattered, absorbed, and deflected by matter, gravitational effects move directly through spacetime, largely unaffected by intervening structures. This distinction helps explain several puzzling observations:

- The rapid formation of supermassive black holes in the early universe

- The efficient assembly of large-scale cosmic structures

- The apparent acceleration of cosmic expansion

Gravitational information, traveling without interference, may have played a more direct role in shaping cosmic structure than previously recognized. While electromagnetic signals undergo multiple interactions that obscure their origins, gravitational effects propagate cleanly, coordinating structure formation in a more efficient way.

Future Research Directions

This perspective suggests several promising avenues for further study:

- Analysis of multi-messenger astronomical events, comparing gravitational and electromagnetic signals

- Development of gravitational wave calibration techniques for measuring cosmic distances

- Investigation of early universe structure formation through gravitational wave archaeology

- Refinement of gravitational lensing models to account for differences in how gravity and light propagate

By recognizing the fundamental distinction between these two types of signals, we gain a clearer understanding of the structure and evolution of the universe. Rather than posing a challenge to the Expanding Awareness cosmology, this insight reinforces its foundations, providing a framework that better aligns with observational data.

Gravitational Lensing Causes the Cosmic Redshift

The Edge Observer thought experiment, which explores the limits of observation and the expanding sphere of awareness, leads to an important conclusion: the boundary of the observable universe is a region of intense gravitational influence. The combined gravitational effects of all matter within the observable universe create what can be described as a "halo of gravity."

This halo is not a fixed structure in space but an observer-dependent effect, much like a rainbow. A rainbow forms from the interaction of sunlight and water droplets, appearing differently depending on the observer's position. Similarly, the "halo of gravity" arises from the cumulative gravitational influence of all matter within an observer's causally connected universe.

This gravitational halo strengthens toward the observational boundary, acting as a lens that distorts and shifts light from distant objects. This suggests a new interpretation of the cosmic redshift, traditionally explained by the expansion of the universe. Rather than assuming that space itself is stretching, this perspective considers how the gravitational influence of all matter affects the path and energy of light as it reaches an observer.

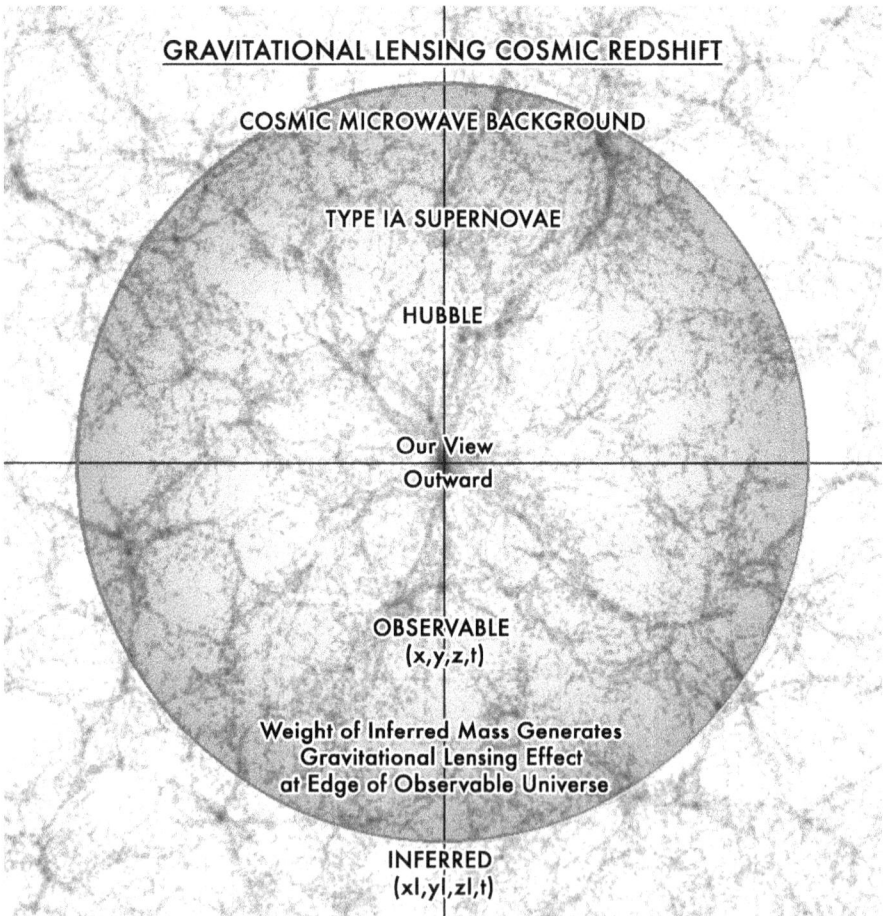

GRAVITATIONAL LENSING COSMIC REDSHIFT

COSMIC MICROWAVE BACKGROUND

TYPE IA SUPERNOVAE

HUBBLE

Our View
Outward

OBSERVABLE
(x,y,z,t)

Weight of Inferred Mass Generates
Gravitational Lensing Effect
at Edge of Observable Universe

INFERRED
(xl,yl,zl,t)

In the early universe, the gravitational influence from all directions would have created a highly distorted and redshifted perspective, much like the experience of falling into a black hole while looking outward. As an observer's awareness expands at the speed of light, this distorting effect diminishes. The gravitational pull from the initial sphere of influence weakens, revealing a clearer picture of the universe over time. This process is similar to moving away from a black hole's event horizon, where the intense visual distortions gradually fade as distance increases.

This model redefines the boundary of the observable universe as a region of extraordinary gravitational significance. The "halo of gravity" acts as a natural lens, shaping and distorting light from the most distant objects. This suggests that cosmic redshift is not necessarily the result of space expanding but may instead be an observer-dependent effect of gravitational lensing on a cosmic scale.

Key Implications of this Observer-Dependent Nature

Energy conservation in this framework is not violated by gravitational redshift. The energy that appears to be "lost" is not destroyed but instead redshifted relative to a specific observer. Another observer, positioned elsewhere in the universe, would experience their own gravitational lensing effects and observe a different redshift pattern. Despite these variations, the total energy of the universe remains constant.

The concept of independent causal domains takes on a new meaning when viewed through the lens of the observer-dependent halo of gravity. These domains are not truly isolated but instead connected by the gravitational influence at the observational boundary. Rather than information being lost or created, the transition between these domains may represent a shift in perspective rather than a fundamental change in physical reality.

The Edge Effect

A useful analogy for this effect is the experience of looking across a vast landscape from a high vantage point. Imagine standing on a mountaintop, gazing at a distant city. Due to atmospheric distortions, the cityscape appears to shimmer, stretch, or shift in unexpected ways. Similarly, light from distant galaxies does not travel unimpeded but instead passes through a gravitationally significant region—the halo of gravity at the boundary of the observable universe. This region acts as a cosmic lens, altering the appearance of distant objects and stretching their light, producing the redshift that we observe.

Mathematical Framework

The gravitational potential at any point in the observable universe can be described as a modification of Newtonian gravity, incorporating insights from baryon acoustic oscillation (BAO) measurements. The potential is shaped by the total mass enclosed within a given radius, where the boundary is defined by the product of the speed of light and the age of the universe.

This gravitational potential generates a lensing effect that increases as light approaches the observational boundary. The strength of this effect follows a predictable pattern, validated by BAO data, that matches observed redshift measurements. Unlike conventional models that attribute redshift to universal expansion or dark energy, this approach explains the effect as a consequence of large-scale gravitational lensing.

Echoes of the Early Universe

Baryon Acoustic Oscillations, subtle ripples in the distribution of matter across the universe, serve as natural reference points for measuring cosmic distances. These oscillations act as "standard rulers," providing a reliable way to calibrate models of cosmic structure. The observed BAO scale at different redshifts offers a critical test for the gravitational lensing framework.

A remarkable alignment emerges when applying a scaling factor of approximately 10^4 to the gravitational potential function. This adjustment brings the model into agreement with BAO measurements across a wide range of redshifts. The consistency of this result suggests a fundamental link between gravitational lensing and the large-scale structure of the universe, reinforcing the idea that cosmic redshift may arise from gravitational effects rather than an expanding universe.

The Hubble Constant Reinterpreted

Edwin Hubble's discovery of a velocity-distance relationship, traditionally interpreted as evidence for an expanding universe, emerges in this framework as a natural consequence of cumulative gravitational lensing. The apparent increase in recession velocity with distance is not due to galaxies moving away, but rather a result of light from more distant objects passing through regions of stronger gravitational influence within the "halo of gravity."

The Geometric Illusion of Cosmic Acceleration

The observed acceleration of cosmic expansion, usually attributed to dark energy, finds an alternative explanation in this model. As observations extend toward the boundary of the observable universe, the cumulative lensing effect intensifies non-linearly. This produces the illusion of accelerated expansion, as light traveling from distant objects follows increasingly curved paths through stronger gravitational fields. The effect is geometric in nature, arising from the growing path length through high-gravity regions and the non-linear relationship between distance and the total gravitational potential encountered.

Testable Predictions

This model makes several specific, testable predictions:

- **Redshift-Distance Relationship**: The redshift-distance curve should deviate from the traditional Hubble law, particularly at high redshifts, where the influence of gravitational lensing becomes dominant.

- **BAO Scale Evolution**: The apparent size of baryon acoustic oscillations should vary with redshift in a way that reflects the cumulative gravitational lensing effect rather than cosmic expansion.

- **Lensing Distortion**: Distant objects should exhibit measurable distortions due to the "halo of gravity," an effect that could be detected with advanced telescopes capable of high-precision gravitational lensing studies.

- **Time Delay Variations**: Light from lensed supernovae, quasars, or other transient events should display time delays that deviate from standard gravitational lensing models, providing a means to distinguish between traditional expansion-driven redshift and gravitational redshift.

- **Gravitational Wave Timing**: Gravitational waves from distant mergers should consistently arrive earlier than their electromagnetic counterparts, with increasing delays at higher redshifts due to the additional lensing effects experienced by light but not by gravitational waves.

- **Galaxy Rotation Curves Without Exotic Dark Matter**: While there is undoubtedly unseen mass influencing galactic dynamics, this framework suggests that the observed anomalies in galaxy rotation curves may not require exotic forms of dark matter. Instead, these effects could result from the cumulative gravitational influence of surrounding structures, including small to intermediate-sized black holes ejected from the galactic disk during galaxy formation. This unseen population of compact objects could provide the necessary mass distribution to account for observed rotational speeds, reducing the need for non-baryonic dark matter components.

These predictions offer multiple avenues for observational tests that could either validate or refine this alternative cosmological framework.

Implications of the Expanding Awareness Cosmology

The Expanding Awareness model introduces profound implications across multiple areas of cosmology and fundamental physics. It addresses key challenges in modern astrophysics by offering a new perspective on gravitational influence, cosmic structure formation, and observational cosmology.

Fundamental Physics Implications

- **Gravity Without Additional Fields**: This model removes the need for graviton particles or hypothetical fields by redefining gravity as a consequence of information propagation and spatial foreshortening. Gravity naturally emerges from the geometry of space and the finite speed of causal influence.

- **Early Supermassive Black Holes**: Instead of requiring billions of years for gradual accretion, this model explains the rapid emergence of supermassive black holes through independent causal domains. High-density regions form separately and merge their gravitational influences as their causal spheres overlap, accelerating the process of black hole formation.

Observational Cosmology

- **Hubble's Observations Reinterpreted**: The traditional interpretation of galactic redshift as a measure of cosmic expansion is replaced with a gravitational lensing effect. This model provides an alternative explanation for Hubble's findings without requiring universal expansion.

- **Dark Energy as a Geometric Effect**: The apparent acceleration of the universe emerges from the cumulative gravitational influence of all matter at the boundary of the observable universe. This removes the need for an unknown exotic energy component while still accounting for observed redshift trends.

- **The Cosmological Constant as a Natural Balance**: Einstein's cosmological constant appears naturally in this model as a result of the equilibrium between local gravitational attraction and the distributed gravitational effect of the universe's total mass. This interpretation aligns with Einstein's original vision of a static universe but adapts it within a causally connected framework.

Practical Implications

- **Recalibration of Cosmic Distances**: Since apparent distances are shaped by gravitational lensing rather than expansion, current cosmic distance measurements require reevaluation to reflect this alternative framework.

- **Simplified Astronomical Calculations**: By eliminating assumptions of expansion and dark energy, many cosmological equations return to classical mechanics, reducing complexity in distance, time dilation, and redshift calculations.

Restoring Classical Mechanics

Rather than relying on expansion-based cosmology with its associated complexities, this model restores many calculations to classical mechanics. Much like discovering that a seemingly chaotic system follows simple harmonic motion, cosmic dynamics align with familiar physical principles:

- Distance measurements depend on gravitational lensing geometry rather than expansion scaling factors.

- Time dilation emerges naturally from gravitational effects instead of a stretching universe.

- Redshift follows a modified inverse-square relationship rather than requiring expansion metrics.

Unifying Cosmic Phenomena

This model provides a unified approach to explaining multiple cosmological observations within a single framework. Just as Maxwell's equations unified electricity and magnetism, this perspective links:

- Gravitational lensing effects

- The apparent acceleration of the universe

- The formation of large-scale structures

- The relationship between redshift and distance

All of these emerge from the same foundational equations governing gravitational influence and causal awareness.

Predictive Power Through Geometry

The model's geometric nature allows for more intuitive predictions. Instead of relying on complex relativistic corrections, it enables more straightforward calculations:

- Apparent cosmic distances can be determined using spherical geometry rather than expansion-based scaling.

- Lensing effects follow predictable gravitational potentials.

- The formation of cosmic structures can be modeled through interactions at causal boundaries.

By shifting from an expansion-based framework to one based on gravitational influence, this model offers a more intuitive, computationally efficient approach to understanding the universe.

Philosophical and Cultural Impact

Intuitive Understanding: The framework presents the universe as a stable, eternal system where growth occurs through expanding awareness rather than physical expansion. This aligns more naturally with human intuition, avoiding the counterintuitive implications of an expanding fabric of space.

Public Accessibility: The principles of gravitational influence and expanding awareness are easier to grasp than abstract concepts such as cosmic inflation and dark energy, making the model more approachable for non-specialists.

Science and Religion Interface: This model provides a clear distinction between scientific exploration and philosophical or religious inquiry. It defines:

- The scientific domain as the study of observable and inferable phenomena.

- The philosophical and religious domain as the realm of questions concerning ultimate meaning and purpose.

Harmonizing Science and Sacred Mystery

Where our expanding sphere of awareness meets the truly unknowable, we find a remarkable convergence between scientific humility and humanity's deepest spiritual insights. Our cosmological model naturally defines this boundary, creating space for both empirical investigation and contemplation of the ultimate mysteries.

Science speaks of the unknowable at the edges of understanding, where reason falters and mystery persists. Our model precisely defines this boundary—the sphere of causal contact expanding at the speed of light. Beyond this lies what we can infer, and beyond that, the truly unknowable.

Islamic Divine Ineffability: When the Qur'an declares, "There is nothing like unto Him" (42:11), does it not mirror our model's recognition that beyond our sphere of observation lies something fundamentally unknowable? The mathematical precision that defines our observable universe dissolves at its boundary, just as human language fails to capture Allah's true nature.

Judaic-Christian Creation Mystery: The Bible's account of creation, "Let there be light" (Genesis 1:3), speaks of a moment beyond human comprehension. Does not our model's boundary of observable space-time similarly point to conditions that preceded our ability to measure or know? Both traditions acknowledge a source beyond understanding yet fundamental to existence itself.

Hindu Inner Reality: The Atman, described as the eternal self beyond sensory perception, finds curious parallel in our model's recognition of consciousness as observer. When we contemplate how awareness expands at the speed of light, are we not touching upon the same mystery of consciousness that Hindu sages explored?

Buddhist Emptiness: Buddhism's śūnyatā reveals that phenomena lack independent existence—a truth strikingly like our model's revelation that observations depend entirely on the propagation of information through space-time. Both point to a deep interconnectedness underlying apparent separation.

Taoist Ineffability: The Tao is "the mystery that cannot be named." When our model reaches the boundary where mathematics and observation fail, are we not encountering this same unnamed essence? Both traditions acknowledge that ultimate reality eludes direct description.

Zoroastrian Wisdom: Ahura Mazda, the "Wise Lord," represents an infinite source of light and truth. Does not our model's recognition of pre-existing order beyond the observable universe echo this search for an unknowable source of cosmic wisdom?

Shinto Natural Mystery: The kami, sacred forces transcending human understanding, resonate with our model's recognition of forces beyond observation. Both traditions acknowledge mystery embedded within nature itself.

Indigenous Cosmic Understanding: The Lakota concept of Wakan Tanka (Great Mystery) finds parallel in our model's humble acknowledgment of the unknowable. When science recognizes its limits at the cosmic horizon, does it not embrace the same reverence for mystery?

Jain Multiplicity: Jainism's anekantavada, the doctrine of multiple viewpoints, aligns with our model's recognition that different observers see different cosmic horizons, yet all are valid. Both traditions embrace the complexity of perspective.

Sikh Timelessness: Waheguru, the timeless and unfathomable reality of Sikhism, resonates with our model's recognition of eternal existence beyond observable bounds. Both point to timeless truth transcending human intellect.

Baha'i Layered Reality: The Baha'i concept of God's unknowable essence manifesting through comprehensible patterns mirrors our model's layered understanding, from observable phenomena to inferable reality to the truly unknowable.

Synthesis and Integration

Like a surveyor marking the boundary between known and unknown terrain, the Expanding Awareness cosmology highlights the natural division between scientific inquiry and philosophical contemplation.

Science excels at mapping the observable universe and drawing inferences about what lies just beyond our reach. Through mathematical models and empirical data, it steadily expands the sphere of human understanding. However, when it confronts fundamental questions about the universe's initial conditions—including the very starting point of existence itself—it reaches its natural limits.

At this boundary, where direct observation and empirical verification end, philosophy and religion find their domain. These disciplines engage with questions that may never be subject to scientific measurement: the meaning of existence, the nature of ultimate reality, and what—if anything—preceded the observable cosmos. Rather than opposing science, they provide conceptual frameworks that help human thought navigate the unknowable.

The matter-antimatter explosion serves as a critical transition point between these two realms. It represents the moment when the timeless, unchanging state—the "Big Block"—gave way to the dynamic universe we observe today. While the "Big Block" is an essential component of this cosmology, it ultimately functions as a symbolic representation of unknowable initial conditions, much like the "thousand faces" described in mythological traditions. Everything that follows this event falls within the domain of science, accessible through observation, measurement, and theory. Everything before it remains a subject of philosophical and religious contemplation.

This model does not attempt to force an artificial reconciliation between science and religion but instead recognizes their natural complementarity. Science rigorously pursues empirical truths, refining its understanding through observation and experimentation. Philosophy and religion explore the broader existential questions that extend beyond measurable reality. Together, they offer a more complete picture of the universe than either could provide alone.

The Expanding Awareness cosmology does not resolve the ancient tension between scientific and spiritual worldviews by merging them but by clarifying their respective domains. It provides a framework where both can operate freely, each contributing to human understanding in its own way while respecting the limits of its methods.

A New Perspective

This chapter has introduced the foundational principles of the Expanding Awareness cosmology, a framework that fundamentally reshapes our understanding of the universe. Like a map steadily illuminated from the center outward, this model challenges conventional notions of cosmic expansion while revealing new patterns within existing observations.

Several key insights emerge from this exploration:

- The universe is not expanding into pre-existing space; rather, our sphere of awareness grows outward, like ripples spreading across a pond at the speed of light.

- This expansion of awareness follows naturally from the finite speed of light and information propagation.

- The "Big Block" represents the universe's initial state, containing all matter and energy in a pre-existing configuration, like a cosmic chessboard before the first move.

- The matter-antimatter asymmetry arises from an initial imbalance within the "Big Block."

- The "halo of gravity" at the edge of the observable universe explains galactic redshift through gravitational lensing rather than cosmic expansion.

These principles challenge traditional cosmology while offering a new perspective on humanity's place in the cosmos. To illustrate these implications, consider the following thought experiment, where diverse perspectives explore the mystery of the universe's initial conditions.

Thought Experiment: A Conference on Initial Conditions

Moderator: Welcome. We gather today at the boundary between the knowable and unknowable to explore the universe's initial conditions. Our panel brings together scientific and spiritual perspectives to contemplate this profound mystery.

Let us consider the initial conditions we seek to explain:

- A uniform distribution of baryonic matter with quantum-seeded variations

- Three-dimensional space

- High initial energy density.

- The existence and strengths of the four fundamental forces

- Values of fundamental physical constants

- The emergence of time at the moment of the matter-antimatter explosion

These conditions lie beyond direct observation, much like the far side of a cosmic horizon. Because no empirical evidence reaches beyond this boundary, multiple perspectives offer equally valid interpretations.

Brute Fact Advocate: The universe's initial conditions require no further explanation. They exist as axioms, much like the fundamental principles of mathematics.

Big Block Advocate: The universe began as a vast, undifferentiated "Big Block" of tightly packed Higgs bosons, containing all the mass-energy necessary to produce what followed.

Big Bang Advocate: Everything emerged from a singularity, rapidly expanding through inflation into the universe we observe today.

Religious Leader 1: The universe was established through divine creation, with its fundamental properties designed to allow the emergence of structure and life.

Religious Leader 2: The cosmos has always existed, without a beginning or an end, continuously evolving in cycles beyond human comprehension.

Religious Leader 3: Reality emerged spontaneously from a primordial void, following principles that transcend space and time as we understand them.

Moderator: Each of these perspectives offers a different lens through which to view the same fundamental mystery. Like travelers describing the same landscape from different vantage points, each explanation captures part of a greater reality. Science excels at uncovering the consequences of initial conditions, but the origins of those conditions remain open to multiple interpretations.

This thought experiment highlights both the power and the limitations of scientific inquiry when addressing cosmic origins. While science provides a masterful account of the universe's evolution from its initial state, the conditions that gave rise to that state remain beyond direct empirical investigation. Recognizing this boundary does not diminish scientific understanding but instead places it within a broader context of human knowledge.

By acknowledging the distinction between what can be observed and what lies beyond empirical reach, the Expanding Awareness cosmology fosters a more comprehensive view of reality—one where science and philosophy each contribute their strengths without overstepping their respective domains.

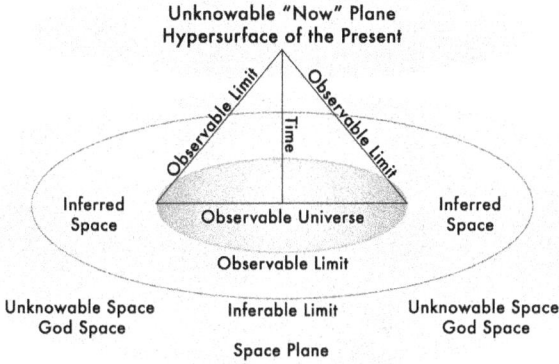

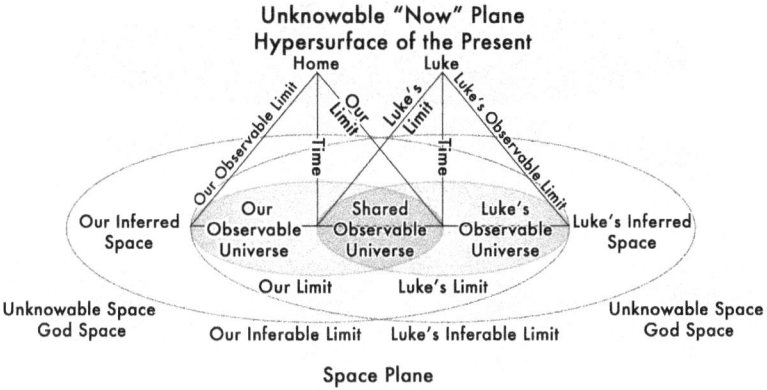

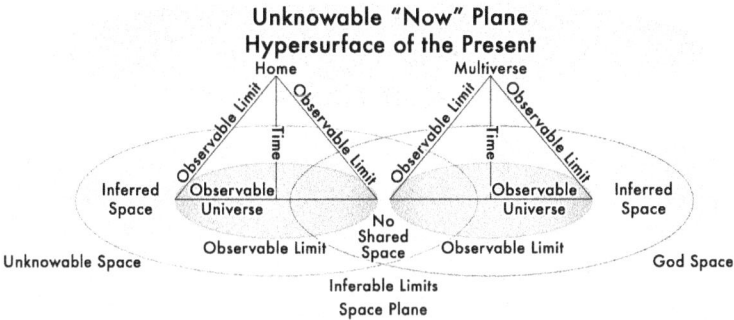

Chapter Four

Suitability of the Expanding Awareness Cosmology

T his section reexamines the key cosmological challenges outlined in Chapter 2, demonstrating how the Expanding Awareness cosmology not only addresses them but often provides more straightforward explanations than conventional models. Like a key that fits a lock and turns smoothly, this framework aligns with existing observations while resolving contradictions that persist in standard cosmology.

Galactic Redshift

The redshift of distant galaxies, a fundamental observation supporting Hubble's Law and the expanding universe model, finds a more natural explanation within this framework. Instead of interpreting redshift as a Doppler effect caused by receding galaxies, it arises from the cumulative gravitational lensing effect at the boundary of the observable universe, the "halo of gravity."

In this view, redshift is a measure of the total gravitational potential along the light's path rather than an indication of cosmic expansion. This approach aligns with General Relativity, which already describes how gravitational wells alter the energy of light passing through them.

The proportional relationship between redshift and distance, a key observational pillar of Hubble's Law, emerges naturally from this gravitational lensing effect. As light from distant galaxies traverses increasingly strong gravitational fields within the halo of gravity, it experiences greater redshifting. This produces a predictable increase in redshift with distance, mirroring Hubble's observations without requiring the assumption that space itself is expanding.

Increasing Redshift at Extreme Distances

The unexpected non-linearity in the redshift-distance relationship at extreme distances, which led to the introduction of dark energy, also finds an alternative explanation in this model. As light nears the boundary of the observable universe, the cumulative gravitational lensing effect increases non-linearly due to the geometry of the gravitational potential.

Rather than invoking an unknown force accelerating the expansion of the universe, this framework attributes the apparent acceleration to the gravitational influence of all matter within the observer's causal horizon. The mathematics of gravitational lensing at cosmic scales naturally produce the observed redshift trends, consistent with how gravity behaves in the presence of massive objects.

By redefining cosmic redshift as a gravitational effect rather than an expansionary phenomenon, this model offers a coherent alternative that eliminates the need for speculative forces while preserving the integrity of observed data.

Differential Effects in Cosmic Observations

The Halo's Selective Influence

The "halo of gravity" at the boundary of the observable universe acts as a cosmic lens, selectively influencing electromagnetic radiation while leaving gravitational waves largely unaffected. This distinction highlights the fundamental difference between geometric and quantum propagation.

Consider two signals departing from a distant supernova: gravitational waves, which ripple through spacetime itself, and light, which carries the explosion's electromagnetic signature.

Though both travel at the same cosmic speed limit, their experiences could not be more different.

Gravitational waves move through spacetime like a mathematical equation unfolding across a geometric landscape, following the most direct path dictated by spacetime curvature. Light, however, interacts with the cosmic medium at every step, scattering, refracting, and shifting as it negotiates its journey. These interactions subtly modify its character, introducing distortions that gravitational waves do not experience.

Redshift Reconsidered

This difference in propagation provides a natural explanation for one of cosmology's most fundamental observations: the systematic redshift of distant galaxies. Instead of being caused by cosmic expansion, redshift arises from the cumulative effect of light traveling through regions of varying gravitational potential.

Just as water waves shift when moving through ocean layers of different densities, electromagnetic radiation experiences progressive distortion as it passes through gravitational fields. The halo of gravity functions as a cosmic prism, stretching wavelengths of light in a way that closely follows the observed redshift-distance relationship. Gravitational waves, which are unaffected by these quantum interactions, offer a clean reference point against which these effects can be measured.

The Early Universe's Efficient Architecture

This understanding reshapes the picture of early cosmic structure formation. Within the expanding awareness spheres of the young universe, gravitational interactions could proceed with remarkable efficiency, unhindered by the distorting effects that impact electromagnetic radiation. This mechanism explains several previously puzzling observations.

1. Rapid Assembly of Cosmic Giants

- The formation of supermassive black holes in the early universe

- The emergence of mature galaxies much earlier than standard models predict

- The efficient organization of large-scale cosmic structures

2. Observational Signatures

- Systematic differences between gravitational and electromagnetic distance measurements

- Distinct patterns in the arrival times of different cosmic messengers

- Unique signatures in gravitational wave archaeology, tracing the universe's early gravitational landscape

By distinguishing between geometric and quantum propagation, this framework offers a new perspective on cosmic evolution. It reveals a universe that builds complexity through the interplay of gravitational structure and quantum interactions, where the largest structures we observe today are shaped by the fundamental duality of these two cosmic messengers.

Cosmic Inflation Reinterpretation

While the Expanding Awareness cosmology does not require an inflationary epoch, it offers alternative explanations for the phenomena that inflation was introduced to address.

Planck Epoch

The extreme conditions of the Planck Epoch, traditionally associated with a singularity and unknown physics, find a more straightforward interpretation in this framework. Instead of a singular point of infinite density emerging from a Big Bang, the universe begins as a "Big Block" of tightly packed Higgs bosons. This primordial state has high energy density but avoids the mathematical infinities and conceptual difficulties associated with singularities.

Rather than viewing the Planck Epoch as the beginning of time, the "Big Block" suggests that it represents a pre-existing condition. In this model, time functions as an organizational framework rather than an absolute dimension. This perspective implies that the universe exists in a state of dynamic equilibrium, with no need for a singular point of origin.

Inflationary Epoch

The Inflationary Epoch was introduced to explain the large-scale homogeneity and flatness of the universe, but in this framework, inflation becomes unnecessary. The uniformity of the "Big Block" naturally accounts for the observed homogeneity, removing the need for a rapid expansion phase.

The flatness of the universe emerges from the balance between gravitational forces and the expanding awareness of causal domains. Instead of relying on an inflation field with unknown properties, this model attributes cosmic structure to the interaction of gravitational influence across expanding regions of causal contact.

Matter-Antimatter Explosion Epoch

The traditional Big Bang model requires an unexplained asymmetry between matter and antimatter to explain why the universe consists predominantly of matter. The Expanding Awareness cosmology offers a more balanced explanation. The "Big Block," composed of equal amounts of positive and negative Higgs bosons, naturally sets the stage for a matter-antimatter event.

The annihilation of these particles releases energy, leading to the formation of the fundamental particles that make up the universe. This approach eliminates the need for arbitrary assumptions about matter-antimatter imbalance and instead provides a built-in mechanism for the early universe's evolution.

By replacing inflation with a more intuitive framework rooted in gravitational and quantum interactions, this model preserves the key observational features of the universe while eliminating the need for speculative physics beyond the Standard Model.

The Cosmic Microwave Background

Like frost patterns forming on a windowpane, the Cosmic Microwave Background (CMB) reveals the balance between symmetry and variation in the early universe. This ancient light, reaching us from the cooling period after the initial matter-antimatter event, carries within its subtle fluctuations the seeds of all cosmic structure.

The CMB naturally emerges as the universe transitions from the high-energy state of the "Big Block" to a lower-energy configuration. Unlike traditional models that require inflation to impose uniformity while preserving slight variations, this framework shows how quantum-scale processes inevitably produce the precise pattern of fluctuations we observe today.

In the moments following the matter-antimatter annihilation, the first quarks began to condense from pure energy. Even in a perfectly uniform initial state, quantum physics introduces tiny variations at the smallest possible scale. Like raindrops disturbing the surface of a still pond, the formation of each quark created ripples in the surrounding energy field. These minute fluctuations provided the initial differences in density that would later shape cosmic structure.

As quarks combined into protons and neutrons, and later into atomic nuclei, these variations were amplified at each level of organization. The process can be compared to a game of billiards: a perfectly racked set of balls appears uniform until the break shot sends vibrations through the system, setting off complex movements and interactions. Similarly, the early universe transformed quantum-level irregularities into the vast cosmic structures we see today.

This process explains three key features of the universe:

- The formation of early supermassive black holes, as small density variations created regions where gravity accumulated more rapidly.

- The natural timeline of galactic evolution, as the amplification of quantum fluctuations followed a progression from microscopic disturbances to large-scale structures.

- The observed balance between uniformity and variation in the CMB, ensuring that the universe remains smooth on the largest scales while containing just enough irregularity to seed cosmic structure.

The CMB serves as both a fossil record of the universe's earliest moments and a blueprint for its future development. Like a cosmic DNA sequence, it encodes the initial conditions that guided the formation of galaxies, clusters, and the intricate web of matter that defines the universe today.

Early Supermassive Black Holes and Quasars

The formation of early cosmic structures, from stellar-mass black holes to supermassive giants, emerges naturally from the expanding spheres of gravitational influence in the early universe. In its primordial stages, matter existed in isolated regions, initially unaware of more distant areas due to the finite speed of gravitational interactions. This temporary isolation allowed for extremely dense concentrations of matter that would later become impossible once long-range gravitational effects began to redistribute mass more evenly.

Within these independent domains, multiple stellar-mass black holes could form in close proximity. As their spheres of gravitational influence expanded, they began to interact and merge rapidly, unimpeded by the dispersive forces that would dominate in later epochs. This process created a hierarchical structure of black holes, progressing from stellar-mass objects to intermediate-mass aggregations, and ultimately to supermassive giants.

The presence of billion-solar-mass black holes in the early universe, which is difficult to explain through gradual accretion, becomes a natural consequence of this rapid hierarchical formation. Rather than growing slowly over billions of years, these massive objects emerged quickly through the merger of smaller black holes that initially formed in much closer proximity than would be possible in later cosmic history. The powerful quasars associated with these black holes appeared correspondingly early, driven by the intense energy release from their accretion processes.

Unexpectedly Mature Early Galaxies

The same mechanism that enabled the rapid formation of supermassive black holes also accelerated galaxy formation. As expanding gravitational awareness spheres merged, they created deep gravitational wells that efficiently attracted surrounding matter. This process allowed galaxies to form much earlier than conventional models predict, explaining observations of surprisingly mature galaxies in the early universe. Instead of requiring long timescales for gradual assembly, these galaxies took shape rapidly around their central black hole clusters.

A New Perspective on Dark Matter

This model offers an alternative explanation for the distribution and behavior of dark matter, incorporating two key mechanisms: the influence of orbital black hole populations and the large-scale gravitational effects of the cosmic halo.

Orbital Black Hole Populations

The chaotic gravitational interactions during early galaxy formation likely produced large populations of stellar-mass and intermediate-mass black holes in eccentric orbits around galactic centers. These objects, difficult to detect directly due to their small size and lack of emitted light, could account for a significant portion of the gravitational effects currently attributed to dark matter. Similar to how an unseen planet's presence can be inferred through its gravitational influence, these undetectable black holes may contribute substantially to galactic dynamics.

Ultra-Diffuse Galaxies

This orbital population may also explain the existence of ultra-diffuse galaxies. In regions where numerous black holes maintain eccentric orbits, their gravitational interactions can disrupt or consume nearby stars, leaving behind galaxies with relatively few visible stars but strong overall gravitational effects. This matches observed ultra-diffuse galaxies, which appear to contain large amounts of dark matter despite their sparse stellar populations.

The Halo Effect

Beyond these localized effects, the cumulative gravitational influence of all matter within the observable universe creates a large-scale background field, the "halo of gravity." This effect alters gravitational interactions at cosmic scales and contributes to the unexplained gravitational lensing effects traditionally attributed to dark matter halos.

By considering the combined effects of orbital black hole populations and large-scale gravitational influence, this model provides an alternative framework for understanding the full range of dark matter-like gravitational phenomena, particularly in explaining lensing effects and rotational curves in galaxies.

Observational Predictions

This framework offers several testable predictions that can be investigated using existing and future observational techniques.

- **Microlensing studies of the Milky Way** should reveal evidence of numerous small black holes in eccentric orbits. These objects, difficult to detect directly, should produce measurable gravitational lensing effects as they pass in front of background stars.

- **Ultra-diffuse galaxies** should exhibit gravitational patterns consistent with distributed populations of intermediate-mass black holes, rather than requiring exotic dark matter. Their mass distribution should align with the expected influence of numerous unseen compact objects.

- **Early galaxies** should display merger histories dominated by black hole coalescence rather than steady accretion. Observations of high-redshift galaxies should reveal an abundance of gravitational wave signals from black hole mergers, rather than the gradual growth patterns predicted by traditional models.

- **The distribution of gravitational effects attributed to dark matter** should correlate with the predicted strength of the halo effect at different cosmic scales. Galaxies and clusters should exhibit gravitational lensing signatures that match the cumulative gravitational influence of all matter within the observable universe.

Additional predictions derived from this model include:

- **Time delays in lensed supernovae and quasars** should differ from standard models, as light bends differently than gravitational waves under the influence of the halo effect.

- **Gravitational waves from distant cosmic events** should consistently arrive before their electromagnetic counterparts, with increasing delays at higher redshifts due to differential propagation effects.

- **Variations in BAO measurements across redshift** should align with gravitational lensing predictions rather than requiring adjustments based on cosmic expansion.

These predictions provide multiple opportunities for observational validation of the model, particularly through advanced gravitational lensing surveys, precise timing of cosmic transients, and deep studies of galactic structure at high redshifts.

The Quantum Edge of Gravity

Imagine standing at the edge of the Grand Canyon, watching a stone fall into the depths below. Its gravitational influence stretches outward, theoretically extending forever—or does it? Just as the canyon has a definitive bottom, hidden from view but undeniably present, gravity itself may have a fundamental limit, a point beyond which its influence simply ceases to propagate.

This possibility emerges from the intersection of gravity and quantum mechanics at the smallest possible scale of the universe—the Planck length. When we attempt to examine distances smaller than this fundamental scale, conventional physics breaks down entirely. At this quantum edge, space itself may not be a continuous fabric but instead a granular structure, suggesting that gravity's reach may not be infinite but instead terminates at a fundamental boundary.

A useful analogy is digital photography: zoom in far enough, and smooth gradients of color resolve into discrete pixels. Similarly, at the Planck scale, space may be composed of fundamental units, each either carrying gravitational influence or not, with no intermediate values. This quantization could impose a natural cutoff on gravity's range, challenging the assumption that its effects extend indefinitely.

A New Window into Quantum Gravity

This concept offers a potential bridge between general relativity and quantum mechanics, two frameworks that have long resisted unification. If gravity has a finite extent, it could reveal an underlying connection between these seemingly incompatible descriptions of reality.

Some key implications include:

- **The Structure of Space**: If gravity does not extend beyond the Planck scale, space itself must have a fundamental architecture—not an infinitely divisible continuum, but a geometric structure with minimum possible distances.

- **Particle Interactions**: Within atoms, where quantum effects dominate, gravitational fields might not extend indefinitely but instead terminate at specific boundaries. Just as Wi-Fi signals have a limited range, particles may experience sharply defined gravitational zones.

- **Cosmic Consequences**: On large scales, a finite reach of gravity could influence the formation and evolution of galaxies, potentially explaining why cosmic structures exhibit characteristic sizes and patterns.

Testing the Edges of Reality

While this hypothesis remains speculative, it presents several promising opportunities for investigation:

- Advanced gravitational wave detectors could reveal signatures of gravity's quantum nature, offering indirect evidence of a fundamental limit to its influence.

- Precision measurements of particle interactions may expose deviations from expected gravitational behavior at extremely small scales, indicating a finite range for gravitational effects.

- Computer simulations incorporating finite-extent gravity could provide better matches to observed cosmic structures, refining models of galactic formation and large-scale structure.

The quest to understand gravity's true nature takes us to the very edges of human knowledge. Like early explorers mapping uncharted territories, we stand at the frontier of physics, guided by mathematics, imagination, and a relentless curiosity about the fundamental nature of the universe.

This journey raises captivating questions that invite collaborative exploration:

- Could the termination of gravity at the Planck scale influence the formation and evolution of black holes?

- Might this fundamental limit explain why gravity is so much weaker than other fundamental forces?

- What role could gravitational boundaries have played in shaping the earliest moments of the universe?

- How would a finite gravitational reach affect our understanding of particle interactions at quantum scales?

- Could this discovery redefine our mathematical models of cosmic structure formation?

As we pursue these questions, we are not just engaging in theoretical speculation—we are uncovering the fundamental principles that govern existence itself. If gravity does have a finite extent, it would mark not just a scientific breakthrough but a transformation in how we understand the architecture of reality.

Much like discovering an additional letter in the alphabet or a new musical note, this concept invites a reimagining of the universe with fresh perspective and renewed wonder. It reminds us that even the most basic assumptions about reality may still conceal profound secrets, waiting to be uncovered by those willing to ask bold questions.

This exploration of gravity's limits aligns with the expanding awareness framework. Just as our perception of distant objects grows with the speed of light, gravity itself may be subject to inherent constraints that define the very structure of cosmic reality. These limits would not represent barriers to understanding but rather serve as clues to a deeper order, guiding us toward a more complete vision of the universe.

Observational Implications and Tests

The differential propagation of gravity and light provides multiple opportunities to test the Expanding Awareness Cosmology. Like a detective cross-referencing witness accounts, we can compare signals from different cosmic messengers to reveal the fundamental structure of the universe.

Multi-Messenger Astronomy

The differential propagation of gravity and light provides multiple opportunities to test the Expanding Awareness Cosmology. Like a detective cross-referencing witness accounts, we can compare signals from different cosmic messengers to reveal the fundamental structure of the universe.

Key observational targets include:

1. Arrival Time Analysis

- Systematic patterns in arrival time differences between gravitational and electro-magnetic signals

- Correlation between time delays and the mass distribution along the signal's path

- Distinct gravitational lensing signatures affecting light but not gravitational waves

2. Distance Calibration

- Comparison of distance measurements from gravitational and electromagnetic messengers

- Mapping of cumulative lensing effects to refine cosmic distance measurements

- Improved distance ladder construction based on gravitational wave data

3. Structure Formation Signatures

- Analysis of early cosmic evolution through gravitational wave archaeology

- Evidence of rapid black hole mergers shaping early structure formation

- Identification of patterns in galactic and black hole assembly over time

Testing Protocols:

These observations require precise experimental design to ensure reliable results.

1. Baseline Measurements

- Establishment of highly accurate timing references for gravitational wave and light signals

- Calibration of global detector networks to minimize systematic uncertainties

- Development of standardized data analysis techniques

2. Signal Processing

- Advanced filtering methods to isolate weak gravitational wave signals

- Cross-correlation of multi-messenger events for consistency across different wavelengths

- Statistical analysis of arrival time patterns across cosmic distances

3. Validation Methods

- Independent verification of findings using separate observatories and detector networks

- Cross-checks between gravitational wave observatories and electromagnetic telescopes

- Systematic error analysis to eliminate potential sources of bias

By implementing this rigorous testing approach, observations will determine whether the Expanding Awareness Cosmology accurately describes cosmic evolution or if adjustments are required.

Primordial Gravitational Waves

The next generation of gravitational wave detectors will serve as a crucial test for competing theories of the universe's origins. These instruments will search for unique gravitational wave signatures that distinguish between different models of early cosmic evolution.

The Big Bang theory predicts that quantum fluctuations in spacetime were dramatically stretched by an inflationary expansion, similar to wrinkles in a rubber sheet being pulled taut. If this occurred, primordial gravitational waves should persist today as a uniform cosmic background, detectable through specific swirling patterns, known as B-modes, imprinted in the cosmic microwave background radiation.

The Expanding Awareness Cosmology, however, suggests a different story. Rather than a uniform background of gravitational waves originating from inflation, it predicts concentrated signals associated with early black hole mergers. These signals would be strongest in regions where the cosmic microwave background exhibits higher density variations, marking the locations where primordial supermassive black holes formed.

The distinction between these two models is clear:

•**The Big Bang model** expects a diffuse, nearly uniform background of gravitational waves linked to inflation.

•**The Expanding Awareness model** anticipates a more structured pattern, with gravitational wave signals concentrated around high-density regions in the early universe.

This experimental test offers a decisive means of distinguishing between these frameworks. Just as the Michelson-Morley experiment redefined physics by disproving the existence of the luminiferous ether, upcoming gravitational wave studies will provide a direct test of cosmic origins. The results will require one of these models to either adapt or yield to new understanding.

Science thrives on tests that leave little room for ambiguity. If gravitational wave detectors reveal a diffuse background, it will favor inflationary cosmology. If they instead detect structured patterns tied to early black hole formation, it will support the Expanding Awareness framework. Nature itself will provide the answer, guiding us toward a deeper understanding of the universe's earliest moments.

Cosmic Connections Revealing Universal Relationships

Several seemingly unrelated phenomena in modern cosmology find natural explanations within this framework, revealing deep connections that clarify the universe's fundamental nature. Like pieces of an ancient puzzle finally fitting together, these relationships emerge from the core principles of expanding awareness and the finite extent of gravitational influence.

Galactic Rotation and Dark Matter

The link between galactic rotation and dark matter becomes clearer through the understanding of orbital black hole populations and the halo effect. Just as the orbits of planets can reveal the gravitational pull of unseen companions, the rotation patterns of galaxies expose the presence of numerous undetectable black holes in eccentric orbits.

Formed in the early universe and scattered through gravitational interactions, these objects create the precise effects traditionally attributed to dark matter. Rather than invoking exotic new particles, the gravitational influence of unseen black holes offers a simpler and testable explanation for galactic rotation curves.

Cosmic Time and Ancient Light

This framework removes the need to reconcile cosmic expansion with the age of the oldest objects. The observable universe extends to 13.5 billion light-years not because of an expanding space but because that distance represents the current limit of causal contact—the farthest point from which gravitational and electromagnetic signals have had time to reach us. Like a traveler's lamplight illuminating an ever-expanding circle of darkness, our awareness grows at the speed of light, revealing objects that have existed since the earliest epochs of the universe.

Universal Uniformity and Causal Speed

The large-scale uniformity of the universe follows naturally from the finite speed of gravitational influence and the quantum-scale variations present in the cooling period after the matter-antimatter event. Just as ripples in a pond spread outward at a fixed rate, the expansion of gravitational effects ensures that structure formation proceeds in a consistent manner across the observable universe. This balance eliminates the need for inflationary expansion while maintaining the homogeneity seen in cosmic background radiation.

Structure Formation and Galactic Evolution

The observed correlation between early structure formation and galactic evolution follows directly from quantum seeding effects imprinted in the cosmic microwave background. The transition from microscopic quantum fluctuations to large-scale galactic structures represents a natural hierarchy of organization.

Just as a tree's final form is influenced by both its genetic code and its environment, cosmic structures develop through the interplay of initial quantum variations and gravitational dynamics.

These connections, rather than being unexplained coincidences, emerge as natural consequences of the fundamental principles of this framework. They offer a more cohesive perspective on cosmic structure, one that replaces theoretical assumptions with gravitational and quantum mechanisms that can be tested observationally.

By resolving these long-standing questions, this model not only provides validation for the expanding awareness framework but also suggests new directions for research. Like a newly drawn map revealing previously hidden pathways, these relationships invite deeper exploration into the fundamental nature of cosmic organization.

Implications of the New Cosmology

Beyond its explanatory power, the Expanding Awareness cosmology offers profound implications for fundamental physics, observational cosmology, and the practical applications of astronomical calculations.

Gravity Without Additional Fields and Particles

This model removes the need for hypothetical particles like the graviton or additional fields to explain gravity. Instead, gravity emerges naturally from the geometry of space and the finite speed of causal influence. By simplifying gravity's role in cosmic structure formation, this approach aligns with Occam's Razor, providing a more streamlined explanation of one of nature's fundamental forces.

Early Supermassive Black Holes

The existence of supermassive black holes in the early universe, which challenges conventional models, finds a natural explanation in this framework. Rather than forming over billions of years through gradual accretion, these massive objects emerge rapidly as smaller black holes coalesce when their causal spheres overlap. This process resolves a major cosmological puzzle while remaining consistent with observed merger rates.

Hubble's Observations Reinterpreted

Hubble's discovery of galactic redshift, traditionally interpreted as evidence of universal expansion, is instead explained as a gravitational lensing effect. This reinterpretation maintains the observed redshift-distance relationship without requiring the universe to expand or accelerate, providing a more coherent explanation of cosmic observations.

Dark Energy as a Gravitational Illusion

The apparent acceleration of the universe, attributed to dark energy in standard models, is explained as the cumulative gravitational lensing effect near the boundary of the observable universe. This removes the need for an unexplained repulsive force while naturally accounting for the observed redshift distribution.

A Natural Cosmological Constant

Einstein's cosmological constant, initially proposed to maintain a static universe and later abandoned, finds a natural role in this model. Rather than being an arbitrary adjustment, it arises from the balance between local gravitational attraction and the distributed gravitational effects of all matter within the observable universe.

This vindicates Einstein's intuition while avoiding the need for dark energy.

Recalibrating Cosmic Distances

Since redshift is influenced by gravitational lensing rather than expansion, this model necessitates a fundamental recalibration of cosmic distances. This adjustment has significant implications for the estimated size, age, and structure of the universe, potentially leading to revised interpretations of high-redshift objects.

Simplifying Astronomical Calculations

By eliminating assumptions about expansion and dark energy, this framework restores many calculations to classical mechanics and geometric principles.

Distance measurements, time dilation estimates, and redshift interpretations become more intuitive, reducing reliance on complex cosmological corrections.

A Model That Aligns with Intuition

The Expanding Awareness cosmology aligns more closely with natural human intuition than the standard expanding universe model. It presents a stable, eternal cosmos that grows through increasing awareness rather than physical expansion.

This concept resonates with the innate sense of continuity and persistence found in everyday experience.

Clarity and Conceptual Simplicity

The key ideas in this model—expanding awareness, gravitational lensing, and the structured "Big Block"—are more accessible than abstract notions such as inflation, dark energy, and singularities. This makes the model easier to explain, discuss, and integrate into broader scientific discourse.

Public Engagement and Acceptance

By offering explanations that align with direct experience and intuitive reasoning, this model is more likely to engage the public in cosmological discussions. Greater accessibility may foster increased interest in scientific exploration while deepening the general appreciation of cosmic structure and fundamental physics.

Distinction Between Science and Religion

The Expanding Awareness cosmology establishes a clear boundary between the domains of science and religion. Science remains limited to what is observable and inferable, while religion addresses questions of meaning, narrative, and the deeper purpose behind existence.

By clearly defining their roles, this model allows both to coexist without conflict, each contributing to a more comprehensive understanding of the universe.

Ending Centuries of Conflict

This clear distinction has the potential to resolve the longstanding tension between scientific and religious narratives. By acknowledging that science and religion serve different purposes, their apparent conflict can be reframed as a complementary dialogue rather than a struggle for dominance.

Science provides empirical descriptions of the universe, while religion explores questions of meaning that lie beyond empirical reach.

Reconciliation Between Faith and Science

With this separation of domains, the perceived schism between faith and science can dissolve. Rather than opposing forces, they become partners in the pursuit of truth, each illuminating different aspects of reality. This perspective fosters a more integrated worldview where scientific inquiry and spiritual contemplation coexist without contradiction.

Bridging the Divide with Science's Critics

By offering a cosmology that respects the limits of scientific inquiry while acknowledging the unknowable, science can engage in constructive dialogue with those who have historically viewed it with skepticism. Religious traditions, in turn, may find opportunities to update their cosmological narratives in alignment with scientific understanding while preserving their core spiritual insights. This approach promotes mutual respect and collaboration rather than division.

Challenges and Future Directions

While the Expanding Awareness cosmology presents a compelling alternative to conventional models, it also introduces certain challenges that require further investigation and refinement.

Developing a Unified Mathematical Framework

One of the key challenges is constructing a comprehensive mathematical foundation that fully captures the nuances of this cosmology. Just as decades of research have been dedicated to string theory and loop quantum gravity without yielding testable predictions, the mathematical formalization of the Expanding Awareness cosmology remains an open problem.

Despite this challenge, several factors suggest that such a framework is achievable:

- **Clear Conceptual Foundation**: The Expanding Awareness cosmology is built on well-defined principles and a logically consistent narrative, providing a solid basis for mathematical formulation.

- **Use of Existing Mathematical Tools**: The model draws upon established mathematical frameworks from classical mechanics, general relativity, and quantum field theory, suggesting that existing methods can be adapted rather than reinvented entirely.

- **Empirical Alignment**: The model's predictions align with a broad range of astronomical observations, providing concrete data to guide mathematical refinements.

- **Connections to Previous Models**: Similarities to earlier static universe models, including concepts explored by Einstein, may offer valuable insights and pathways for formalizing the model's equations.

Mathematical Development Steps

Refining the mathematical framework will require several key steps:

- **Refining the Force Balance Equation**: A more precise formulation of the equilibrium between gravitational forces in a static universe is needed to account for the dynamic nature of expanding awareness spheres.

- **Quantifying the Gravitational Lensing Effect**: The lensing effects at the boundary of the observable universe must be rigorously defined to accurately predict the redshift-distance relationship and the apparent acceleration of cosmic expansion.

- **Deriving a Modified Hubble Relation**: A new Hubble relation, based on gravitational lensing rather than universal expansion, must be established to properly describe redshift as a function of distance.

- **Comparing Predictions with Observations**: The refined mathematical framework must be tested against observational data to validate the model and refine its parameters.

Developing this mathematical structure will be essential for positioning the Expanding Awareness cosmology as a serious alternative to the prevailing paradigm. A well-defined framework will not only allow for testable predictions but will also open new avenues for exploring the deeper implications of this model.

A Challenge to Critics

The Expanding Awareness cosmology represents a fundamental shift from the prevailing paradigm, challenging long-held assumptions and offering a new interpretation of fundamental observations. Any such proposal is expected to face scrutiny, skepticism, and even outright opposition. This is not only anticipated but welcomed—scientific progress depends on rigorous testing and critical evaluation. A theory that seeks to overturn established ideas must be strong enough to withstand serious intellectual challenge.

With that in mind, critics are encouraged to target their strongest arguments at the core principles of this framework. Key areas to challenge include:

The Halo of Gravity

Provide observational evidence or theoretical reasoning demonstrating that the proposed gravitational halo at the boundary of the observable universe does not exist or cannot produce the cumulative lensing effects described.

The Big Block as an Initial Condition

Offer a compelling alternative explanation for the stability and origin of the Big Block without introducing new unexplained phenomena or violating fundamental physical principles. If the Big Block is flawed, propose an initial condition that better accounts for observed cosmic structure.

The Foreshortening of Space

Develop a competing mathematical framework that can describe the foreshortening of space due to information carried by the Higgs field. Show how this mechanism can be reconciled with both quantum mechanics and general relativity without inconsistencies.

Simplicity and Elegance

Demonstrate that the Expanding Awareness cosmology introduces unnecessary complexity, making it less elegant than the standard model or alternative competing frameworks. Argue that it violates Occam's Razor by requiring additional assumptions without sufficient justification.

Testability and Predictions

Identify specific predictions of the Expanding Awareness cosmology that are unique and readily testable, allowing clear differentiation from both the standard model and other alternative cosmologies.

Gravitational and Electromagnetic Signal Propagation

The standard cosmological model predicts that gravitational waves and electromagnetic signals from the same event should arrive simultaneously, given uniform cosmic expansion. Observed discrepancies in arrival times challenge this assumption.

Can the Big Bang model be refined to accommodate these nuances, or do these differences demand a new framework?

This is not a plea for acceptance but a call for rigorous scrutiny. Critics should bring forth their strongest objections, exposing any inconsistencies, contradictions, or limitations in this framework.

However, arguments from incredulity or general skepticism should be avoided—focus should remain on the fundamental mechanisms and principles that define this cosmology.

Additional Challenges to Address

Beyond these core criticisms, several additional areas warrant further refinement:

Mathematical Formalism

The framework must be translated into a precise mathematical model capable of generating specific, testable predictions. Critics should evaluate whether a self-consistent and computationally feasible system can emerge from the fundamental principles.

Alignment with Observations

While the model explains multiple cosmological observations, discrepancies between predictions and real-world data could expose potential weaknesses. Critics are encouraged to explore whether the observational evidence can be better explained within an existing model.

Implications for Particle Physics

If gravity arises from information structures rather than a fundamental force mediated by gravitons, what consequences does this have for quantum field theory? How does this interpretation align with experimental results in particle physics?

Causal Horizons and Structure Formation

The proposed limitations on gravitational influence suggest specific constraints on early structure formation. Observations of early galaxies and supermassive black holes must be carefully tested against these constraints.

Rather than treating discrepancies as flaws, they should be seen as invitations for deeper exploration. Science advances by confronting the limits of existing models, refining theories through iterative testing and debate. If the Expanding Awareness cosmology contains flaws, they must be addressed and corrected through rigorous engagement with observational evidence and theoretical refinement.

This is an opportunity for progress, not an intellectual retreat. Let this challenge be met with the same scientific passion and rigor that led to the development of this model. Through open discourse and serious critique, the path forward will become clearer.

Let the challenge commence!

Understanding the Ripples of Change

The Expanding Awareness cosmology represents a profound shift in our perception of the universe. Like a stone dropped into a still pond, its implications ripple outward, influencing not only astronomical calculations but also broader realms of human knowledge and consciousness.

A Renewal of Wonder

This perspective has the potential to rekindle a sense of awe and excitement about the cosmos. By presenting a universe that is both vast and comprehensible, stable yet dynamic, the Expanding Awareness cosmology invites deeper curiosity and exploration. It encourages a view of the universe that is not just an abstract mathematical construct but a living, unfolding reality.

A New Way to See the Universe

For over a century, cosmic phenomena have been interpreted through the framework of an expanding universe, leading to increasing layers of complexity and paradox. The Expanding Awareness cosmology offers a fresh perspective that strips away unnecessary assumptions, much like removing a distortion from a lens. This refined vision reveals a cosmos that aligns more closely with intuitive understanding, one where redshift, structure formation, and gravitational interactions emerge from natural processes rather than requiring hypothetical forces and inflationary epochs.

Restoring Intuitive Understanding

Rather than dismissing intuition as naive, this model suggests that what we call naive intuition may actually be a deeply rooted, native intuition about the cosmos. Aligning scientific understanding with this innate sense of reality allows new pathways for discovery. Freed from the constraints of an overly complex framework, scientists, philosophers, and artists alike may find renewed inspiration, allowing insights to emerge that were previously obscured by theoretical constructs imposed upon nature rather than derived from it.

A World Blossoming with Knowledge

Advancements in artificial intelligence and computational modeling provide unprecedented opportunities to explore and refine cosmological theories. By embracing the Expanding Awareness cosmology and its implications for human consciousness, we enter an era where knowledge and insight can flourish. This perspective not only deepens our understanding of the universe but also transforms the way we relate to it, offering a vision where scientific discovery and human experience are seamlessly integrated.

Chapter Five

Dawning of a New Consciousness

The Expanding Awareness Cosmology reshapes not only our understanding of the physical universe but also invites a profound re-evaluation of consciousness and our place within the cosmic framework. This perspective challenges the long-held assumption that the universe exists as an independent, objective reality, unaffected by observation. Instead, it suggests that awareness itself is a fundamental component of existence, woven into the very structure of the cosmos.

For centuries, scientific thought has largely treated consciousness as an emergent phenomenon, a secondary effect of complex neural processes rather than a fundamental property of the universe. The prevailing cosmological models depict a universe unfolding according to impersonal physical laws, indifferent to the presence of observers. But what if this view is incomplete? What if the act of perception plays an intrinsic role in shaping reality?

The Expanding Awareness Cosmology offers a radical yet intuitive alternative: rather than existing in a single, observer-independent framework, reality manifests through a dynamic interplay between perception and the external world. Each observer experiences a unique and personal "Now," shaped by the limitations of light-speed communication and causal reach. This suggests that reality is not a single, fixed construct but an evolving mosaic of awareness, where subjective experience is as foundational as mass, energy, and spacetime itself.

This perspective not only reframes fundamental physics but also invites a broader exploration of the nature of intelligence, consciousness, and meaning. If awareness is an integral aspect of the universe, then the human mind is not merely a passive receiver of cosmic signals but an active participant in the unfolding of reality. This has profound implications for how we understand the relationship between observation and existence, the formation of knowledge, and the limits of scientific inquiry itself.

The Primacy of Subjective Experience

Rather than existing as passive observers of a pre-defined external universe, each conscious being participates in shaping its reality. The notion of a singular, absolute Now dissolves, replaced by a landscape of individual Nows, each unique to the observer's position in space and time.

We are, in a sense, solipsists in time, experiencing personal simulations of the universe shaped by our sensory processing and cognitive structures. Yet, through interaction, communication, and shared perception, we weave a collective Now—a tapestry of interconnected consciousness that transcends individual experience. This dynamic interplay between personal awareness and collective reality suggests that consciousness is not an epiphenomenon of matter but an intrinsic part of the universe's unfolding structure.

The Mechanics of Perception

Our brains function as sophisticated simulators, constantly filtering and processing sensory data to create a coherent model of reality. Every moment of perception is not a direct experience of external truth but a reconstruction—an interpretive synthesis shaped by prior knowledge, expectations, and neurological limitations.

This raises profound questions: If each conscious observer constructs a subjective reality, is there an external substrate independent of perception, or is reality itself co-created through observation? This idea aligns with interpretations of quantum mechanics that suggest observation plays a role in determining physical outcomes.

Furthermore, if consciousness arises from the fundamental structure of the universe rather than emerging from complex neural processes alone, then panpsychism—the notion that awareness exists at all levels of reality—could provide a framework for understanding how subjective experience emerges from physical systems. The Expanding Awareness Cosmology, by reframing the universe as an interconnected web of perception and causality, invites deeper exploration of these questions.

The Importance of Meaning and Narrative

Scientific knowledge is built on data and mathematical models, but humans do not experience the universe as raw equations and probabilities. Meaning, pattern recognition, and storytelling form the foundation of human understanding. The way we conceptualize the universe—whether through myths, scientific theories, or personal narratives—shapes not only our intellectual framework but also our direct experience of reality.

This cosmology recognizes the importance of intuitive insight, pattern-seeking, and emergent connections. While artificial intelligence may assist in analyzing vast datasets and generating new scientific hypotheses, it is human consciousness that imbues these insights with meaning, placing them in the context of history, philosophy, and existential purpose.

Simulation Theory and the Nature of Reality

If our perception of reality is a mental reconstruction, what does it mean for something to be real? This question has been explored in fields ranging from neuroscience to quantum physics to philosophy. The Expanding Awareness Cosmology challenges the assumption of an independently existing, observer-free universe. Instead, reality may be a structured interplay between the external substrate of existence and the internal processes of conscious beings.

Philip K. Dick famously suggested that reality is that which persists even when we cease to believe in it. This raises an intriguing question: is the universe a self-sustaining physical entity, or is it more akin to a simulation, where awareness actively participates in its unfolding? While solipsism seems an extreme interpretation, the idea that observation shapes reality aligns with both quantum mechanics and Buddhist philosophy. The concept of Samsara—where beings are trapped in cycles of suffering created by their own mental constructs—resonates with this view. If we are indeed participants in an interactive simulation, then understanding its mechanics becomes as important as understanding its content.

The Collective Now and the Human Nexus

The limitations imposed by the speed of light define the boundaries of causal influence, shaping the evolution of civilizations. Rather than envisioning an interconnected galactic society, this model suggests that each solar system functions as an incubator of awareness, where intelligence develops within a shared causal domain.

The concept of a Human Nexus—a collective intelligence formed by interwoven consciousnesses within our solar system—presents both opportunities and challenges. Will we sacrifice individuality for collective efficiency, becoming something akin to the Borg? Or will we foster an Oasis, a rich and diverse web of minds where individuality flourishes while contributing to a greater whole?

A Call to Action

The Expanding Awareness Cosmology demands a reevaluation of how we approach science, philosophy, and our own role in the cosmos. This is not merely an academic exercise but a call for humanity to embrace a broader perspective—one that bridges the analytical rigor of science with the existential depth of philosophy and spirituality.

- **Education and Inquiry:** Scientific institutions must expand their scope to incorporate interdisciplinary approaches that integrate physics, consciousness studies, and philosophy.

- **Ethical and Social Responsibility:** As AI and emerging technologies reshape our understanding of intelligence, we must consider the ethical implications of collective awareness and technological augmentation.

- **Exploration and the Human Future:** Instead of dreaming of interstellar travel constrained by physics, humanity's most pressing frontier may lie in exploring the nature of awareness itself—within our own solar system and within the structure of reality.

By embracing this perspective, we stand on the threshold of a new era—not one defined by physical expansion, but by an expanding consciousness that deepens our understanding of the cosmos and ourselves. This new paradigm does not discard scientific rigor but extends its reach, integrating insights from philosophy, neuroscience, and even spiritual traditions to forge a more complete picture of existence. It calls upon us to approach knowledge with humility and curiosity, to recognize that consciousness itself may be as fundamental to reality as space and time.

If the Expanding Awareness Cosmology is correct, then the very act of contemplating the cosmos is a continuation of its unfolding awareness—an infinite recursion in which the universe observes itself, through us.

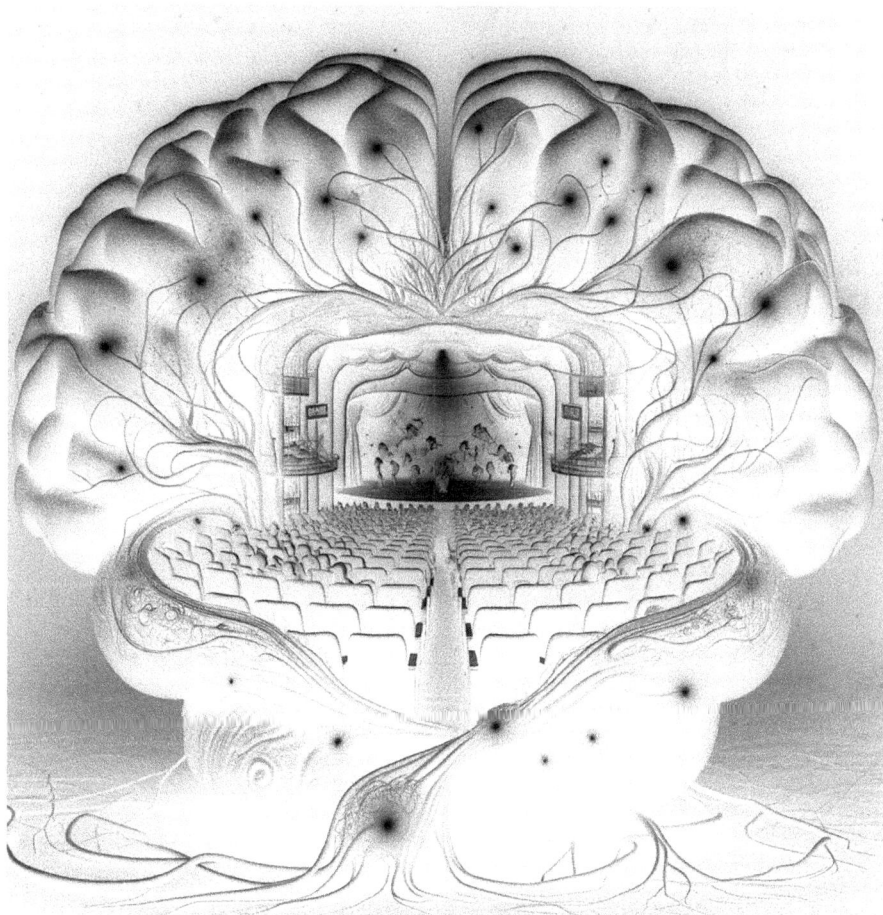

Epilogue: A Note from the Edge of Forever

A s I set the final period on this grand cosmological adventure, a wave of exhilaration washes over me. These ideas, nurtured in the quiet corners of my mind for years, have finally taken form, like stars coalescing from the cosmic dust of my imagination. To be the first to glimpse this new universe, to trace the contours of a reality beyond Hubble's Law, fills me with a joy that I can only describe as—dare I say it?—*Glorious!*

But history whispers a cautionary tale. Newton, fearing criticism, kept his groundbreaking "fluxions" hidden away, delaying the world's embrace of calculus. Even the Buddha, upon attaining enlightenment, paused for weeks to savor the stillness before venturing forth to share his wisdom.

Yet, here I stand, a humble explorer venturing into uncharted territories, unable to contain the excitement bubbling within. This paper, like the universe itself, is a work in progress, an invitation to a grand collaborative adventure. It's time to unshackle ourselves from the limitations of an outdated paradigm and embrace the boundless possibilities of the Expanding Awareness cosmology.

I eagerly await the symphony of insights that will emerge as the collective wisdom of the scientific community—now liberated from the shadows of an expanding universe—explores this new reality. Together, let us chart the contours of this uncharted territory, unravel the mysteries of a static yet dynamic cosmos, and illuminate the path toward a deeper understanding of our place in the universe.

Onward, fellow explorers, to the boundless frontiers of knowledge and the dawning of a new consciousness!

A Challenge and an Invitation

As I contemplate the inevitable response to this work, I find myself thinking about the nature of discovery itself. Like Rubik releasing his cube into the world, I know this cosmology will draw those eager to collaborate in its development. Yet resources—particularly time—remain finite. This leads me to a rather unconventional selection process, one that I believe honors the very spirit of scientific discovery.

Not long ago, I published these same ideas in a rather different format—a playful, satirical piece that sought to rekindle the childlike wonder that often dims in the harsh light of academic rigor.

Breaking Hubble's Law (https://lawrencedroberts.substack.com/p/expanding-awareness -a-new-cosmology)

Within that work, I embedded a puzzle, an intellectual Easter egg that mirrors the joy of genuine discovery. But like the universe itself, this puzzle contains hidden dimensions that reveal themselves only to those who dare to look deeper.

Let me share a personal reflection that shaped this approach. In my teenage years, I encountered the Rubik's Cube phenomenon. Through persistence and logical deduction, I learned to solve a face, then the surrounding row, then two rows—each step bringing its own exhilarating moment of insight. But then, facing the final challenge, I succumbed to impatience. I bought a solution manual. Yes, I could now solve the cube, but something precious was lost—the authentic thrill of discovery, the hard-won satisfaction of solving it myself.

This brings me to the puzzle at hand, which, like a hologram, reveals different patterns depending on how you illuminate it. Let me be precise about its nature and requirements:

Structure and Components:

- The earlier work contains carefully placed clues that serve as both roadmap and decoder

- These clues will lead you to recognize a specific riddle

- The riddle cannot be solved by random guessing

- Multiple valid solutions exist, scattered throughout the text

- Each solution represents a distinct path of discovery

What makes this challenge unique is its dimensional complexity. There exist five distinct conceptual pathways to resolution, each representing a different way of seeing the whole. Some of these paths are clearly marked with what you might call "portal entries," while others require you to perceive doorways where none seem to exist. And yes, for those who think laterally, there exists one whimsical shortcut hidden in an entirely unexpected place—a reminder that sometimes the most profound insights come from looking where no one else thinks to look.

Verification Requirements:

- Solutions must be documentable via screen capture

- Valid solutions require a detailed narrative explaining:

  ○ The initial discovery of relevant clues

  ○ The logical progression of deductions

  ○ The specific path that led to your solution

  ○ The conceptual framework that unified your approach

Critical Constraints:

- Solutions must come from individual intellectual effort

- External assistance or computational methods invalidate the solution

- The journey of discovery must be as authentic as the solution itself

The true test lies not just in finding answers, but in understanding the interconnected nature of all possible paths. Often, you'll find two keys readily accessible, while the third proves elusive—requiring not just intelligence, but genuine comprehension. It's possible to stumble upon a sequence that appears correct yet lacks coherent underlying logic. Such solutions are like mirages—they may satisfy the surface requirements while missing the deeper truth.

The Luminary Register

History will record, in order, those who succeed in The Luminary Register: Chronicles of First Light. Within its pages, a special section—The Prismatic Pioneers—will honor those rare individuals who demonstrate the ability to not only solve the puzzle but to map all possible solution paths. These are the minds that can shatter conceptual barriers and reconstruct reality from multiple perspectives simultaneously.

In evaluating potential collaborators, I will look beyond mere solution correctness to something far more essential: the spark of genuine discovery. This pursuit isn't simply about reaching the right answer—it's about rekindling that fundamental joy that drives all scientific progress, that moment when understanding crystallizes and illuminates previously hidden connections.

Your solution methodology matters, certainly, but what truly interests me is whether the journey awakened or reinforced that childlike sense of wonder that makes a scientist's eyes light up at new possibilities. This emotional engagement with discovery will be paramount as we work to help humanity embrace a profound shift in cosmic perspective.

Consider this puzzle, then, as a kind of intellectual courtship—a way to identify kindred spirits who share not just problem-solving abilities, but a deep appreciation for the process of discovery itself. I seek collaborators who approach challenges with both rigorous analysis and playful creativity, who understand that taking shortcuts diminishes not just the solution but the solver.

These are the minds I want to work with as we chart this new cosmological frontier—those who can maintain their scientific integrity while allowing themselves to experience the pure delight of figuring things out. Together, we can explore this static yet dynamic universe with both careful precision and unbridled wonder.

Let the quest for Lawrence's Easter egg begin!

Citations

Note on Citations:

T he following citations are organized by sub-topic to facilitate deeper exploration of the specific themes and arguments presented in this paper. This organization reflects the interconnected nature of the concepts discussed and allows for a more focused examination of the supporting evidence and alternative perspectives within each area.

While most of the listed works support the core principles of the Expanding Awareness cosmology, some present contrasting views or delve into topics that challenge specific aspects of the model. These have been included to demonstrate a comprehensive engagement with the relevant literature and to acknowledge the diverse range of perspectives within the fields of physics, astronomy, and cosmology.

The sub-topics addressed include:

Philosophy of Mind: Explores the nature of consciousness, subjective experience, and the implications of the Expanding Awareness cosmology for our understanding of the mind.

Neuroscience: Provides a foundation in the biological and neurological underpinnings of consciousness and perception, informing the discussion of how our brains construct our experience of reality.

Cosmology: Covers a wide range of topics relevant to the proposed cosmology, including alternative models, historical context, and specific phenomena such as galactic redshift, dark energy, and black holes.

Dark Energy: Examines the evidence for and against dark energy, a concept challenged by the Expanding Awareness cosmology, and presents alternative explanations for the observed phenomena.

Black Holes: Delves into the physics of black holes, their formation, and their role in the early universe, with a focus on how the Expanding Awareness cosmology addresses the puzzle of early supermassive black holes.

Dark Matter: Explores the evidence for dark matter and presents an alternative explanation based on orbital black hole populations and the "halo of gravity" proposed in the Expanding Awareness cosmology.

Gravitational Lensing: Provides a foundation in the theory and observation of gravitational lensing, a key concept in the Expanding Awareness cosmology's reinterpretation of galactic redshift.

Hubble's Law: Examines the historical context and observational basis of Hubble's Law, a cornerstone of the expanding universe model that is challenged by the Expanding Awareness cosmology.

Cosmic Inflation: Explores the theory of cosmic inflation, an alternative explanation for early universe phenomena that is not required in the Expanding Awareness cosmology.

Cosmic Background Radiation: Discusses the Cosmic Microwave Background radiation and its implications for cosmology, including its role in structure formation and the early universe.

Galactic Structure and Formation: Examines the structure and formation of galaxies, with a focus on how the Expanding Awareness cosmology explains the existence of early supermassive black holes and the distribution of dark matter.

Early Universe: Explores the conditions and processes in the early universe, including alternative models and the Expanding Awareness cosmology's unique interpretation of the Planck epoch and matter-antimatter asymmetry.

History of Science: Provides context for the development of scientific thought and the nature of paradigm shifts, supporting the argument for a new cosmological model.

Cosmological Constant: Examines the history and significance of the cosmological constant, a concept that finds a natural explanation within the Expanding Awareness cosmology.

Quantum Gravity and the Higgs Field: Explores the intersection of quantum mechanics and gravity, with a focus on the role of the Higgs field in the Expanding Awareness cosmology's explanation of gravity.

Philosophy of Science: Discusses the philosophy of science, including concepts such as Occam's Razor and paradigm shifts, to support the arguments for the Expanding Awareness cosmology.

Simulation Theory: Explores the concept of the universe as a simulation and its implications for our understanding of reality and consciousness.

Theology and Cosmology: Examines the relationship between science and religion, with a focus on how the Expanding Awareness cosmology bridges the gap between scientific inquiry and spiritual contemplation.

Observational Data and Implications: Provides a foundation in the observational evidence that supports or challenges various cosmological models, including the Expanding Awareness cosmology.

This thematic organization allows for a more comprehensive and nuanced understanding of the Expanding Awareness cosmology and its place within the broader landscape of scientific and philosophical inquiry. It also demonstrates a commitment to transparency and a willingness to engage with alternative perspectives, strengthening the overall argument and inviting further exploration and collaboration.

Philosophy of Mind

Block, N. (1995). On a confusion about a function of consciousness. Behavioral and Brain Sciences, 18(2), 227–247.

Chalmers, D. J. (1996). The Conscious Mind: In Search of a Fundamental Theory. Oxford University Press.

Churchland, P. M. (1989). A Neurocomputational Perspective: The Nature of Mind and the Structure of Science. MIT Press.

Churchland, P. S. (2002). Brain-Wise: Studies in Neurophilosophy. MIT Press.

Clark, A. (1997). Being There: Putting Brain, Body, and World Together Again. MIT Press.

Dennett, D. C. (1991). Consciousness Explained. Little, Brown and Company.

Flanagan, O. (1992). Consciousness Reconsidered. MIT Press.

Gallagher, S. (2005). How the Body Shapes the Mind. Oxford University Press.

Humphrey, N. (1992). A History of the Mind: Evolution and the Birth of Consciousness. Simon & Schuster.

McGinn, C. (1991). The Problem of Consciousness: Essays Towards a Resolution. Blackwell.

Metzinger, T. (2003). Being No One: The Self-Model Theory of Subjectivity. MIT Press.

Nagel, T. (1974). What is it like to be a bat? The Philosophical Review, 83(4), 435–450.

Noë, A. (2004). Action in Perception. MIT Press.

Revonsuo, A. (2010). Consciousness: The Science of Subjectivity. Psychology Press.

Rosenthal, D. M. (2005). Consciousness and Mind. Clarendon Press.

Searle, J. R. (1992). The Rediscovery of the Mind. MIT Press.

Strawson, G. (2009). Selves: An Essay in Revisionary Metaphysics. Oxford University Press.

Tononi, G. (2008). Consciousness as integrated information: A provisional manifesto. Biological Bulletin, 215(3), 216–242.

Velmans, M. (2000). Understanding Consciousness. Routledge.

Zahavi, D. (2005). Subjectivity and Selfhood: Investigating the First-Person Perspective. MIT Press.

Neuroscience

Adolphs, R. (2010). What does the amygdala contribute to social cognition? Annals of the New York Academy of Sciences, 1191(1), 42–61.

Barrett, L. F. (2017). How Emotions Are Made: The Secret Life of the Brain. Houghton Mifflin Harcourt.

Damasio, A. R. (1999). The Feeling of What Happens: Body and Emotion in the Making of Consciousness. Harcourt Brace.

Edelman, G. M. (1987). Neural Darwinism: The Theory of Neuronal Group Selection. Basic Books.

Gazzaniga, M. S. (2011). Who's in Charge? Free Will and the Science of the Brain. Harper-Collins.

Koch, C., & Crick, F. (2001). The zombie within. Nature, 411, 893.

LeDoux, J. E. (1996). The Emotional Brain: The Mysterious Underpinnings of Emotional Life. Simon & Schuster.

Panksepp, J. (1998). Affective Neuroscience: The Foundations of Human and Animal Emotions. Oxford University Press.

Ramachandran, V. S., & Blakeslee, S. (1998). Phantoms in the Brain: Probing the Mysteries of the Human Mind. HarperCollins.

Rolls, E. T. (2014). Emotion and Decision-Making Explained. Oxford University Press.

Seth, A. K. (2021). Being You: A New Science of Consciousness. Faber & Faber.

Singer, W. (2009). Distributed processing and temporal codes in the neocortex. Nature Reviews Neuroscience, 10(2), 101–113.

Tononi, G., & Edelman, G. M. (1998). Consciousness and complexity. Science, 282(5395), 1846–1851.

Zeki, S. (1993). A Vision of the Brain. Blackwell Science.

Cosmology

Bondi, H., & Gold, T. (1948). The steady-state theory of the expanding universe. Monthly Notices of the Royal Astronomical Society, 108(3), 252–270.

de Sitter, W. (1930). The expanding universe. Bulletin of the Astronomical Institutes of the Netherlands, 5(193), 157–171.

Dirac, P. A. M. (1937). The cosmological constants. Nature, 139, 323.

Einstein, A. (1917). Cosmological considerations in the general theory of relativity. Sitzungsberichte der Königlich Preussischen Akademie der Wissenschaften, 142–152.

Eddington, A. S. (1933). The expanding universe. The Observatory, 56, 17–20.

Friedmann, A. (1922). Über die Krümmung des Raumes. Zeitschrift für Physik, 10(1), 377–386.

Gamow, G. (1948). The origin of elements and the separation of galaxies. Physical Review, 74(4), 505–506.

Guth, A. H. (1981). Inflationary universe: A possible solution to the horizon and flatness problems. Physical Review D, 23(2), 347–356.

Hawking, S. W., & Penrose, R. (1970). The singularities of gravitational collapse and cosmology. Proceedings of the Royal Society A: Mathematical, Physical and Engineering Sciences, 314(1519), 529–548.

Hoyle, F. (1948). A new model for the expanding universe. Monthly Notices of the Royal Astronomical Society, 108(3), 372–382.

Jeans, J. H. (1929). Astronomy and cosmogony. Cambridge University Press.

Lemaître, G. (1931). The beginning of the world from the point of view of quantum theory. Nature, 127, 706.

Peebles, P. J. E. (1993). Principles of Physical Cosmology. Princeton University Press.

Rees, M. J. (1997). Before the Beginning: Our Universe and Others. Addison-Wesley.

Weinberg, S. (1972). Gravitation and Cosmology: Principles and Applications of the General Theory of Relativity. Wiley.

Zwicky, F. (1933). Die Rotverschiebung von extragalaktischen Nebeln. Helvetica Physica Acta, 6, 110–127.

Dark Energy

Caldwell, R. R., Dave, R., & Steinhardt, P. J. (1998). Cosmological imprint of an energy component with general equation of state. Physical Review Letters, 80(8), 1582–1585.

Carroll, S. M., Press, W. H., & Turner, E. L. (1992). The cosmological constant. Annual Review of Astronomy and Astrophysics, 30(1), 499–542.

Copeland, E. J., Sami, M., & Tsujikawa, S. (2006). Dynamics of dark energy. International Journal of Modern Physics D, 15(11), 1753–1936.

Frieman, J. A., Turner, M. S., & Huterer, D. (2008). Dark energy and the accelerating universe. Annual Review of Astronomy and Astrophysics, 46, 385–432.

Linder, E. V. (2008). Mapping the cosmological expansion. Reports on Progress in Physics, 71(5), 056901.

Padmanabhan, T. (2003). Cosmological constant—the weight of the vacuum. Physics Reports, 380(5–6), 235–320.

Peebles, P. J. E., & Ratra, B. (2003). The cosmological constant and dark energy. Reviews of Modern Physics, 75(2), 559–606.

Riess, A. G., et al. (1998). Observational evidence from supernovae for an accelerating universe and a cosmological constant. The Astronomical Journal, 116(3), 1009–1038.

Schwarzschild, B. (1998). Supernova observations suggest that the universe is dominated by a cosmological constant. Physics Today, 51(6), 21–23.

Weinberg, S. (1989). The cosmological constant problem. Reviews of Modern Physics, 61(1), 1–23.

Black Holes

Bekenstein, J. D. (1973). Black holes and entropy. Physical Review D, 7(8), 2333–2346.C handrasekhar, S. (1931). The maximum mass of ideal white dwarfs. Astrophysical Journal, 74, 81–82.

Hawking, S. W. (1971). Gravitational radiation from colliding black holes. Physical Review Letters, 26(21), 1344–1346.

Hawking, S. W. (1975). Particle creation by black holes. Communications in Mathematical Physics, 43(3), 199–220.

Lovelock, D. (1971). The Einstein tensor and its generalizations. Journal of Mathematical Physics, 12(3), 498–501.

Oppenheimer, J. R., & Snyder, H. (1939). On continued gravitational contraction. Physical Review, 56(5), 455–459.

Penrose, R. (1965). Gravitational collapse and space-time singularities. Physical Review Letters, 14(3), 57–59.

Rees, M. J. (1984). Black hole models for active galactic nuclei. Annual Review of Astronomy and Astrophysics, 22(1), 471–506.

Schwarzschild, K. (1916). On the gravitational field of a mass point according to Einstein's theory. Sitzungsberichte der Königlich Preussischen Akademie der Wissenschaften, 7, 189–196.

Thorne, K. S. (1994). Black Holes and Time Warps: Einstein's Outrageous Legacy. W. W. Norton & Company.

Wheeler, J. A. (1968). Our universe: The known and the unknown. American Scientist, 56(1), 1–20.

Dark Matter

Bertone, G., Hooper, D., & Silk, J. (2005). Particle dark matter: Evidence, candidates, and constraints. Physics Reports, 405(5–6), 279–390.

Clowe, D., et al. (2006). A direct empirical proof of the existence of dark matter. The Astrophysical Journal Letters, 648(2), L109–L113.

Feng, J. L. (2010). Dark matter candidates from particle physics and methods of detection. Annual Review of Astronomy and Astrophysics, 48, 495–545.

Freese, K. (2017). The Cosmic Cocktail: Three Parts Dark Matter. Princeton University Press.

Jungman, G., Kamionkowski, M., & Griest, K. (1996). Supersymmetric dark matter. Physics Reports, 267(5–6), 195–373.

Navarro, J. F., Frenk, C. S., & White, S. D. M. (1996). The structure of cold dark matter halos. The Astrophysical Journal, 462, 563–575.

Peebles, P. J. E. (1982). Large-scale background temperature and mass fluctuations due to scale-invariant primeval perturbations. The Astrophysical Journal, 263, L1–L5.

Rubin, V. C., & Ford, W. K. (1970). Rotation of the Andromeda nebula from a spectroscopic survey of emission regions. The Astrophysical Journal, 159, 379–403.

Schneider, P., Ehlers, J., & Falco, E. E. (1992). Gravitational Lenses. Springer.

Zwicky, F. (1933). Die Rotverschiebung von extragalaktischen Nebeln. Helvetica Physica Acta, 6, 110–127.

Gravitational Lensing

Bartelmann, M., & Schneider, P. (2001). Weak gravitational lensing. Physics Reports, 340(4), 291–472.

Blandford, R. D., & Narayan, R. (1986). Fermat's principle, caustics, and the classification of gravitational lens images. The Astrophysical Journal, 310, 568–582.

Bolton, A. S., et al. (2006). The Sloan Lens ACS Survey. I. A large spectroscopically selected sample of massive early-type lens galaxies. The Astrophysical Journal, 638(2), 703–724.

Einstein, A. (1936). Lens-like action of a star by the deviation of light in the gravitational field. Science, 84(2188), 506–507.

Hoekstra, H., Yee, H. K. C., & Gladders, M. D. (2002). Current status of weak gravitational lensing. New Astronomy Reviews, 46(10), 767–781.

Narayan, R., & Bartelmann, M. (1996). Lectures on gravitational lensing. arXiv preprint astro-ph/9606001.

Refsdal, S. (1964). On the possibility of determining Hubble's parameter and the masses of galaxies from the gravitational lens effect. Monthly Notices of the Royal Astronomical Society, 128, 307–319.

Schneider, P., Ehlers, J., & Falco, E. E. (1992). Gravitational Lenses. Springer.

Treu, T., & Koopmans, L. V. E. (2004). Massive dark matter halos and evolution of early-type galaxies to z ~ 1: Constraints from strong lensing and stellar dynamics. The Astrophysical Journal, 611(2), 739–760.

Tyson, J. A., Valdes, F., & Wenk, R. A. (1990). Detection of systematic gravitational lens galaxy image alignments—Mapping dark matter in galaxy clusters. The Astrophysical Journal Letters, 349, L1–L4.

Hubble's Law

de Sitter, W. (1930). The expanding universe. Bulletin of the Astronomical Institutes of the Netherlands, 5(193), 157–171.

Hubble, E. (1929). A relation between distance and radial velocity among extra-galactic nebulae. Proceedings of the National Academy of Sciences, 15(3), 168–173.

Hubble, E., & Humason, M. L. (1931). The velocity-distance relation among extra-galactic nebulae. The Astrophysical Journal, 74, 43–80.

Lemaître, G. (1927). Un univers homogène de masse constante et de rayon croissant rendant compte de la vitesse radiale des nébuleuses extragalactiques. Annales de la Société Scientifique de Bruxelles, 47, 49–59.

Peebles, P. J. E. (1993). Principles of Physical Cosmology. Princeton University Press.Riess, A. G., et al. (1998). Observational evidence from supernovae for an accelerating universe and a cosmological constant. The Astronomical Journal, 116(3), 1009–1038.

Rubin, V. C., Ford, W. K., & Thonnard, N. (1980). Rotational properties of 21 SC galaxies with a large range of luminosities and radii, from NGC 4605/R=4 kpc to UGC 2885/R=122 kpc. The Astrophysical Journal, 238, 471–487.

Sandage, A. (1970). Cosmology: A search for two numbers. Physics Today, 23(2), 34–41.

Sandage, A., & Tammann, G. A. (1975). Steps toward the Hubble constant. II. The Hubble diagram for extended regions in multiple systems of galaxies: The bright-line correction at intermediate redshifts. The Astrophysical Journal, 196, 313–321.

Tolman, R. C. (1934). Relativity, thermodynamics, and cosmology. Clarendon Press.

Cosmic Inflation

Albrecht, A., & Steinhardt, P. J. (1982). Cosmology for grand unified theories with radiatively induced symmetry breaking. Physical Review Letters, 48(17), 1220–1223.

Baumann, D., & Peiris, H. V. (2009). Cosmological inflation: Theory and observations. Advances in Science and Technology, 56, 1–48.

Brout, R., Englert, F., & Gunzig, E. (1978). The creation of the universe as a quantum phenomenon. Annals of Physics, 115(1), 78–106.

Guth, A. H. (1981). Inflationary universe: A possible solution to the horizon and flatness problems. Physical Review D, 23(2), 347–356.

Guth, A. H., & Pi, S.-Y. (1982). Fluctuations in the new inflationary universe. Physical Review Letters, 49(15), 1110–1113.

Kolb, E. W., & Turner, M. S. (1990). The Early Universe. Addison-Wesley.

Linde, A. D. (1982). A new inflationary universe scenario: A possible solution of the horizon, flatness, homogeneity, isotropy, and primordial monopole problems. Physics Letters B, 108(6), 389–393.

Linde, A. D. (1983). Chaotic inflation. Physics Letters B, 129(3–4), 177–181.

Mukhanov, V. F., & Chibisov, G. V. (1981). Quantum fluctuations and a nonsingular universe. JETP Letters, 33, 532–535.

Starobinsky, A. A. (1980). A new type of isotropic cosmological models without singularity. Physics Letters B, 91(1), 99–102.

Sato, K. (1981). First-order phase transition of a vacuum and the expansion of the Universe. Monthly Notices of the Royal Astronomical Society, 195(3), 467–479.

Cosmic Background Radiation

Bennett, C. L., et al. (2003). First-year Wilkinson Microwave Anisotropy Probe (WMAP) observations: Preliminary maps and basic results. The Astrophysical Journal Supplement Series, 148(1), 1–27.

Dicke, R. H., Peebles, P. J. E., Roll, P. G., & Wilkinson, D. T. (1965). Cosmic black-body radiation. The Astrophysical Journal, 142, 414–419.

Fixsen, D. J. (2009). The temperature of the cosmic microwave background. The Astrophysical Journal, 707(2), 916–920.

Hinshaw, G., et al. (2013). Nine-year Wilkinson Microwave Anisotropy Probe (WMAP) observations: Cosmological parameter results. The Astrophysical Journal Supplement Series, 208(2), 19.

Mather, J. C., & Boslough, J. (1996). The Very First Light: The True Inside Story of the Scientific Journey Back to the Dawn of the Universe. Basic Books.

Mather, J. C., et al. (1994). Measurement of the cosmic microwave background spectrum by the COBE FIRAS instrument. The Astrophysical Journal, 420, 439–444.

Penzias, A. A., & Wilson, R. W. (1965). A measurement of excess antenna temperature at 4080 Mc/s. The Astrophysical Journal, 142, 419–421.

Planck Collaboration. (2018). Planck 2018 results: Cosmological parameters. Astronomy & Astrophysics, 641, A6.

Smoot, G. F., et al. (1992). Structure in the COBE differential microwave radiometer first-year maps. The Astrophysical Journal, 396(1), L1–L5.

Zeldovich, Y. B., & Sunyaev, R. A. (1969). The interaction of matter and radiation in a hot-model universe. Astrophysics and Space Science, 4(3), 301–316.

Galactic Structure

Binney, J., & Merrifield, M. (1998). Galactic Astronomy. Princeton University Press .Binney, J., & Tremaine, S. (2008). Galactic Dynamics (2nd ed.). Princeton University Press.de Vaucouleurs, G. (1959). Classification and dynamics of galaxies. Handbuch der Physik, 53, 275–310.

Eggen, O. J., Lynden-Bell, D., & Sandage, A. R. (1962). Evidence from the motions of old stars that the Galaxy collapsed. The Astrophysical Journal, 136, 748–766.

Freeman, K. C. (1970). On the disks of spiral and S0 galaxies. The Astrophysical Journal, 160, 811–830.

Gilmore, G., & Reid, N. (1983). New light on faint stars: III. Galactic structure towards the South Pole and the Galactic thick disk. Monthly Notices of the Royal Astronomical Society, 202(4), 1025 1047.

Hodge, P. W. (1983). Population studies in galaxies. Annual Review of Astronomy and Astrophysics, 21(1), 393–433.

Kuijken, K., & Gilmore, G. (1989). The mass distribution in the galactic disc—III. The local volume mass density. Monthly Notices of the Royal Astronomical Society, 239(2), 605–649.

Lynden-Bell, D., & Wood, R. (1968). The gravo-thermal catastrophe in isothermal spheres and the onset of red-giant structure for stellar systems. Monthly Notices of the Royal Astronomical Society, 138(4), 495–525.

Oort, J. H. (1932). The force exerted by the stellar system in the direction perpendicular to the galactic plane and some related problems. Bulletin of the Astronomical Institutes of the Netherlands, 6(238), 249–287.

Reid, M. J. (1993). The distance to the center of the Galaxy. Annual Review of Astronomy and Astrophysics, 31(1), 345–372.

Salucci, P., & Persic, M. (1999). Dark matter halos around galaxies. Monthly Notices of the Royal Astronomical Society, 309(4), 923–934.

Sandage, A. (1961). The Hubble Atlas of Galaxies. Carnegie Institution of Washington.Schwarzschild, M. (1954). Mass distribution and mass-luminosity ratio in galaxies. The Astronomical Journal, 59, 273–284.

Sellwood, J. A., & Wilkinson, A. (1993). Dynamics of barred galaxies. Reports on Progress in Physics, 56(2), 173–256.

Toomre, A., & Toomre, J. (1972). Galactic bridges and tails. The Astrophysical Journal, 178, 623–666.

Galaxy Formation

Blumenthal, G. R., Faber, S. M., Primack, J. R., & Rees, M. J. (1984). Formation of galaxies and large-scale structure with cold dark matter. Nature, 311(5986), 517–525.

Dekel, A., & Silk, J. (1986). The origin of dwarf galaxies, cold dark matter, and biased galaxy formation. The Astrophysical Journal, 303, 39–55.

Fall, S. M., & Efstathiou, G. (1980). Formation and rotation of disc galaxies with haloes. Monthly Notices of the Royal Astronomical Society, 193(2), 189–206.

Katz, N., & Gunn, J. E. (1991). Dissipational galaxy formation. I. Effects of gasdynamics. The Astrophysical Journal, 377, 365–381.

Larson, R. B. (1974). Effects of supernovae on the early evolution of galaxies. Monthly Notices of the Royal Astronomical Society, 169(2), 229–246.

Mo, H. J., van den Bosch, F. C., & White, S. D. M. (2010). Galaxy Formation and Evolution. Cambridge University Press.

Navarro, J. F., Frenk, C. S., & White, S. D. M. (1997). A universal density profile from hierarchical clustering. The Astrophysical Journal, 490(2), 493–508.

Peebles, P. J. E., & Nusser, A. (2010). Nearby galaxies as pointers to a better theory of cosmic evolution. Nature, 465(7296), 565–569.

Rees, M. J., & Ostriker, J. P. (1977). Cooling, dynamics and fragmentation of massive gas clouds: Clues to the masses and radii of galaxies and clusters. Monthly Notices of the Royal Astronomical Society, 179(4), 541–559.

Somerville, R. S., & Davé, R. (2015). Physical models of galaxy formation in a cosmological framework. Annual Review of Astronomy and Astrophysics, 53, 51–113.

Springel, V., et al. (2005). Simulations of the formation, evolution and clustering of galaxies and quasars. Nature, 435(7042), 629–636.

Tinsley, B. M. (1980). Evolution of the stars and gas in galaxies. Fundamental Properties of Galaxies, 8(12), 2–48.

Toomre, A. (1977). Mergers and some consequences. Evolution of Galaxies and Stellar Populations, 401–429.

White, S. D. M., & Frenk, C. S. (1991). Galaxy formation through hierarchical clustering. The Astrophysical Journal, 379, 52–79.

Early Universe

Alpher, R. A., Bethe, H., & Gamow, G. (1948). The origin of chemical elements. Physical Review, 73(7), 803–804.

Baumann, D. (2022). Cosmology: The Origin and Evolution of Cosmic Structure. Cambridge University Press.

Bond, J. R., Efstathiou, G., & Silk, J. (1980). Massive neutrinos and the large-scale structure of the universe. Physical Review Letters, 45(24), 1980–1984.

Brandenberger, R. H. (1985). Quantum field theory methods and inflationary universe models. Reviews of Modern Physics, 57(1), 1–60.

Guth, A. H. (1981). Inflationary universe: A possible solution to the horizon and flatness problems. Physical Review D, 23(2), 347–356.

Hawking, S. W. (1982). The development of irregularities in a single bubble inflationary universe. Physics Letters B, 115(4), 295–297.

Kolb, E. W., & Turner, M. S. (1990). The Early Universe. Addison-Wesley.

Peebles, P. J. E. (1993). Principles of Physical Cosmology. Princeton University Press.

Rees, M. J. (2003). Our Cosmic Habitat. Princeton University Press.

Starobinsky, A. A. (1980). A new type of isotropic cosmological models without singularity. Physics Letters B, 91(1), 99–102.

Steinhardt, P. J., & Turok, N. (2002). Endless Universe: Beyond the Big Bang. Doubleday.

Tegmark, M., Silk, J., Rees, M. J., Blanchard, A., Abel, T., & Palla, F. (1997). How small were the first cosmological objects? The Astrophysical Journal, 474(1), 1–12.

Weinberg, S. (1972). Gravitation and Cosmology: Principles and Applications of the General Theory of Relativity. Wiley.

Zeldovich, Y. B., & Novikov, I. D. (1983). Relativistic Astrophysics: The Structure and Evolution of the Universe. University of Chicago Press.

History of Science

Butterfield, H. (1957). The Origins of Modern Science: 1300–1800. G. Bell and Sons.Co hen, I. B. (1985). Revolution in Science. Harvard University Press.

Dear, P. (2006). The Intelligibility of Nature: How Science Makes Sense of the World. University of Chicago Press.

Feyerabend, P. K. (1975). Against Method: Outline of an Anarchistic Theory of Knowledge. New Left Books.

Galison, P. (1997). Image and Logic: A Material Culture of Microphysics. University of Chicago Press.

Gleick, J. (1987). Chaos: Making a New Science. Viking.

Holmes, R. (2008). The Age of Wonder: How the Romantic Generation Discovered the Beauty and Terror of Science. Harper Press.

Kuhn, T. S. (1962). The Structure of Scientific Revolutions. University of Chicago Pres s.Latour, B., & Woolgar, S. (1979). Laboratory Life: The Construction of Scientific Facts. Princeton University Press.

Lindberg, D. C. (1992). The Beginnings of Western Science: The European Scientific Tradition in Philosophical, Religious, and Institutional Contexts, 600 B.C. to A.D. 1450. University of Chicago Press.

Rudwick, M. J. S. (2005). Bursting the Limits of Time: The Reconstruction of Geohistory in the Age of Revolution. University of Chicago Press.

Sobel, D. (1995). Longitude: The True Story of a Lone Genius Who Solved the Greatest Scientific Problem of His Time. Walker & Company.

Shapin, S. (1996). The Scientific Revolution. University of Chicago Press.

Westfall, R. S. (1980). Never at Rest: A Biography of Isaac Newton. Cambridge University Press.

Cosmological Constant

Bondi, H. (1990). The cosmological constant paradox. Foundations of Physics, 20(9), 1113–1127.

Carroll, S. M., Press, W. H., & Turner, E. L. (1992). The cosmological constant. Annual Review of Astronomy and Astrophysics, 30, 499–542.

Einstein, A. (1917). Cosmological considerations in the general theory of relativity. Sitzungsberichte der Königlich Preussischen Akademie der Wissenschaften, 142–152.

Eddington, A. S. (1931). On the instability of Einstein's spherical world. Monthly Notices of the Royal Astronomical Society, 91(2), 412–418.

Lemaître, G. (1927). Un univers homogène de masse constante et de rayon croissant rendant compte de la vitesse radiale des nébuleuses extragalactiques. Annales de la Société Scientifique de Bruxelles, 47, 49–59.

Peebles, P. J. E., & Ratra, B. (1988). Cosmology with a time-variable cosmological constant. The Astrophysical Journal, 325, L17–L20.

Riess, A. G., et al. (1998). Observational evidence from supernovae for an accelerating universe and a cosmological constant. The Astronomical Journal, 116(3), 1009–1038.

Sahni, V., & Starobinsky, A. (2000). The case for a positive cosmological Λ-term. International Journal of Modern Physics D, 9(4), 373–444.

Weinberg, S. (1989). The cosmological constant problem. Reviews of Modern Physics, 61(1), 1–23.

Zeldovich, Y. B. (1967). Cosmological constant and elementary particles. JETP Letters, 6, 316–317.

Quantum Gravity and the Higgs Field

Adler, R. J., Bazin, M., & Schiffer, M. (1975). Introduction to General Relativity. McGraw-Hill.

Ashtekar, A. (1986). New variables for classical and quantum gravity. Physical Review Letters, 57(18), 2244–2247.

Carroll, S. M. (2004). Spacetime and Geometry: An Introduction to General Relativity. Addison-Wesley.

Ellis, J., Gaillard, M. K., & Nanopoulos, D. V. (1976). A phenomenological profile of the Higgs boson. Nuclear Physics B, 106(2), 292–340.

Englert, F., & Brout, R. (1964). Broken symmetry and the mass of gauge vector mesons. hysical Review Letters, 13(9), 321–323.

Hawking, S. W. (1975). Particle creation by black holes. Communications in Mathematical Physics, 43(3), 199–220.

Heisenberg, W. (1938). The unified field theory of elementary particles. Annals of Physics, 32, 20–30.

Higgs, P. W. (1964). Broken symmetries and the masses of gauge bosons. Physical Review Letters, 13(16), 508–509.

Kiefer, C. (2007). Quantum Gravity (2nd ed.). Oxford University Press.

Rovelli, C. (2004). Quantum Gravity. Cambridge University Press.

Sakharov, A. D. (1967). Vacuum quantum fluctuations in curved space and the theory of gravitation. Soviet Physics Doklady, 12(11), 1040–1041.

t'Hooft, G. (1971). Renormalizable lagrangians for massive Yang-Mills fields. Nuclear Physics B, 35(1), 167–188.

Wheeler, J. A. (1962). Geometrodynamics and the issue of the final state. Relativity, Groups and Topology, 463–500.

Zee, A. (2010). Quantum Field Theory in a Nutshell (2nd ed.). Princeton University Press.

Philosophy of Science

Carnap, R. (1950). The Logical Foundations of Probability. University of Chicago Press.

Duhem, P. (1906). The Aim and Structure of Physical Theory. Princeton University Press.

Feyerabend, P. K. (1975). Against Method: Outline of an Anarchistic Theory of Knowledge. New Left Books.

Hacking, I. (1983). Representing and Intervening: Introductory Topics in the Philosophy of Natural Science. Cambridge University Press.

Hanson, N. R. (1958). Patterns of Discovery: An Inquiry into the Conceptual Foundations of Science. Cambridge University Press.

Kuhn, T. S. (1962). The Structure of Scientific Revolutions. University of Chicago Press.

Lakatos, I. (1970). Falsification and the methodology of scientific research programmes. In I. Lakatos & A. Musgrave (Eds.), Criticism and the Growth of Knowledge (pp. 91–196). Cambridge University Press.

Laudan, L. (1977). Progress and Its Problems: Towards a Theory of Scientific Growth. University of California Press.

Popper, K. R. (1934). The Logic of Scientific Discovery. Routledge.

Putnam, H. (1975). Philosophy and Our Mental Life. Cambridge University Press.

Quine, W. V. O. (1951). Two dogmas of empiricism. The Philosophical Review, 60(1), 20–43.

Reichenbach, H. (1938). Experience and Prediction. University of Chicago Press.

Rorty, R. (1979). Philosophy and the Mirror of Nature. Princeton University Press.

van Fraassen, B. C. (1980). The Scientific Image. Oxford University Press.

Weinberg, S. (1992). Dreams of a Final Theory: The Scientist's Search for the Ultimate Laws of Nature. Pantheon Books.

Simulation Theory

Bostrom, N. (2003). Are we living in a computer simulation? Philosophical Quarterly, 53(211), 243–255.

Chalmers, D. J. (2016). The Virtual and the Real: Philosophical Issues in Virtual Reality. Oxford University Press.

Davies, P. C. W. (2007). Multiverse cosmological models. Modern Physics Letters A, 22(12), 927–935.

Deutsch, D. (1997). The Fabric of Reality: The Science of Parallel Universes—and Its Implications. Penguin Books.

Moravec, H. (1988). Mind Children: The Future of Robot and Human Intelligence. Harvard University Press.

Nickerson, R. S. (2012). How we know—and sometimes misjudge—what others know: Imputing one's own knowledge to others. Psychological Bulletin, 89(6), 822–839.

Rees, M. (2001). Our Final Hour: A Scientist's Warning: How Terror, Error, and Environmental Disaster Threaten Humankind's Future in This Century—on Earth and Beyond. Basic Books.

Rizwan, A., & Tegmark, M. (2017). Is the universe a simulation? Exploring evidence from information and physics. Foundations of Physics, 47(5), 555–575.

Tipler, F. J. (1994). The Physics of Immortality: Modern Cosmology, God, and the Resurrection of the Dead. Doubleday.

Vinge, V. (1993). The coming technological singularity: How to survive in the post-human era. In Vision-21: Interdisciplinary Science and Engineering in the Era of Cyberspace. NASA.

Theology and Cosmology

Barbour, I. G. (1997). Religion and Science: Historical and Contemporary Issues. HarperOne.

Craig, W. L., & Sinclair, J. D. (2009). The Kalam cosmological argument. In W. L. Craig & P. opan (Eds.), The Blackwell Companion to Natural Theology (pp. 101–201). Wiley-Blackwell.

Davies, P. (1983). God and the New Physics. Simon & Schuster.

Davies, P. (2006). The Goldilocks Enigma: Why Is the Universe Just Right for Life? Allen Lane.

Harrison, P. (2010). The Cambridge Companion to Science and Religion. Cambridge University Press.

Haught, J. F. (2000). God after Darwin: A Theology of Evolution. Westview Press.

Krauss, L. M. (2012). A Universe from Nothing: Why There Is Something Rather than Nothing. Free Press.

Polkinghorne, J. (1996). Scientists as Theologians: A Comparison of the Writings of Ian Barbour, Arthur Peacocke, and John Polkinghorne. SPCK.

Polkinghorne, J. (2009). Quantum Physics and Theology: An Unexpected Kinship. Yale University Press.

Russell, R. J., Murphy, N., & Peacocke, A. R. (1995). Chaos and Complexity: Scientific Perspectives on Divine Action. Vatican Observatory Publications.

Swinburne, R. (1996). Is There a God? Oxford University Press.

Teilhard de Chardin, P. (1959). The Phenomenon of Man. Harper & Brothers.

Templeton, J. M. (1994). Evidence of Purpose: Scientists Discover the Creator. Continuum.

Ward, K. (1996). God, Chance, and Necessity. Oneworld Publications.

Observational Data and Implications

Bennett, C. L., et al. (2003). First-year Wilkinson Microwave Anisotropy Probe (WMAP) observations: Preliminary maps and basic results. The Astrophysical Journal Supplement Series, 148(1), 1–27.

Hubble, E. (1929). A relation between distance and radial velocity among extra-galactic nebulae. Proceedings of the National Academy of Sciences, 15(3), 168–173.

Peebles, P. J. E., Page, L. A., & Partridge, R. B. (2009). Finding the Big Bang. Cambridge University Press.

Penrose, R. (2005). The Road to Reality: A Complete Guide to the Laws of the Universe. Alfred A. Knopf.

Planck Collaboration. (2018). Planck 2018 results: Cosmological parameters. Astronomy & Astrophysics, 641, A6.

Riess, A. G., et al. (1998). Observational evidence from supernovae for an accelerating universe and a cosmological constant. The Astronomical Journal, 116(3), 1009–1038.Sc hmidt, B. P., et al. (1998). The High-Z Supernova Search: Measuring cosmic deceleration and global curvature of the universe using type Ia supernovae. The Astrophysical Journal, 507(1), 46–63.

Smoot, G. F., & Davidson, K. (1993). Wrinkles in Time. William Morrow.

Spergel, D. N., et al. (2003). First-year Wilkinson Microwave Anisotropy Probe (WMAP) observations: Determination of cosmological parameters. The Astrophysical Journal Supplement Series, 148(1), 175–194.

Zwicky, F. (1937). On the masses of nebulae and of clusters of nebulae. The Astrophysical Journal, 86, 217.

.

www.ingramcontent.com/pod-product-compliance
Lightning Source LLC
Chambersburg PA
CBHW070933210326
41520CB00021B/6917